Shortwave Listening on the Road

Other books by Andrew Yoder

Pirate Radio Stations: Tuning in to Underground Broadcasts

The Complete Shortwave Listener's Handbook–4th edition

Build Your Own Shortwave Antennas–2nd edition

Auto Audio: Choosing, Installing, Maintaining, and Repairing Car Stereo Systems

Pirate Radio: The Incredible Saga of America's Underground, Illegal Broadcasters

Shortwave Listening on the Road

The World Traveler's Guide

Andrew Yoder

TAB Books
Division of McGraw-Hill
New York San Francisco Washington, D.C.
Auckland Bogotá Caracas Lisbon London Madrid Mexico City
Milan Montreal New Delhi San Juan Singapore Sydney Tokyo Toronto

McGraw-Hill

A Division of The ***McGraw-Hill*** *Companies*

Published by TAB Books, a division of The McGraw-Hill Companies

pbk 1 2 3 4 5 6 7 8 9 FGR/FGR 9 0 0 9 8 7 6 5

Library of Congress Cataloging-in-Publication Data
Yoder, Andrew R.
Shortwave listening on the road : the world traveler's guide / by Andrew Yoder.
p. cm.
Includes index.
ISBN 0-07-076509-X (p)
1. Shortwave radio—Receivers and reception—Amateurs' manuals. 2. Portable radios—Amateurs' manuals. I. Title.
TK9956.Y642 1995
621.384'151—dc20 95-35151
CIP

Acquisitions editor: Roland Phelps
Editorial team: Joanne Slike, Executive Editor
Susan W. Kagey, Book Editor
Joann Woy, Indexer
Production team: Katherine G. Brown, Director
Ollie Harmon, Coding
Wanda S. Ditch, Desktop Operator
Janice Ridenour, Computer Artist
Design team: Jaclyn J. Boone, Designer
Katherine Lukaszewicz, Associate Designer

076509X
TAB3

To my family, who has tolerated
my sometimes cumbersome attempts at
portable and shortwave listening over the years.

Contents

Appendices

Acknowledgments

Special thanks to: Bill Early of Trans World Radio, Frederica Dochinoiu of Radio Romania International, Ursula Fleck-Jerwin and Sarah Hartley Edwards of Deutsche Welle, Simon Spanswick of the BBC, Aida Hamza of UAE Radio from Abu Dhabi, Graham Barclay of KIWI Radio, Ken McHarg of HCJB, Pilar Salvador of Radio Exterior de Espana, Ronald Grunig of Swiss Radio International, Wafa Ghawi of Radio Damascus, Richard W. Jones and Richard Dentici of WEWN, Alden Forrester and Adrian Peterson of Adventist World Radio, Ashraful Alam of Radio Bangladesh, Michael Reuter of KJES, Arie Schellaars of Radio Australia, Mike Osborne of KNLS, Charlotte Addler of Radio Sweden, Tina Hammers of WSHB, Joe Vincent of Jolly Roger Radio, Helga Dingova of Radio Slovakia International, Levon Ananikian of the Voice of America, Karen Zeck of Trans World Radio—Guam, Jim Moncrief of Grundig, Georgia Morgan of Drake, Patrick McDonald and Chris Lobdell of the Boston Area DXers.

Preface

There's an information revolution occurring. Today, millions of different people and organizations are on a network where the flow of ideas, culture, and information is the goal. You can link up to this network and receive information about raves directly from England, about hot mineral springs from Japan, and about the political unrest in Rwanda and Burundi.

People the world over are discussing the benefits and detractions of the information revolution, and the mass media is embracing the concept of the information highway. It's the new frontier and, like the Old West, it will help to shape the future of the United States and the rest of the world.

Technology is changing rapidly. Just a decade ago, you could only access this network from a house (for the most part). These days, you can even travel and link up to the network with a box of technology as small as your hand. Just plug it in and let the information download.

It isn't the Internet; it's shortwave radio. Shortwave? Aren't they those huge, complicated rigs that old guys talk to each other with? Not anymore! Shortwave radio has always been exciting, but now it's portable, too. You can catch up-to-the-minute headline news from England; grunge, pop, rap, hip-hop, and heavy metal music from Kuwait; communist rhetoric from North Korea; music with a *real* beat from Togo; Rush Limbaugh and other right-wing talk-show hosts from the United States; Christianity from the peaks of the Andes; and much more.

If you want to join in with other forums and talk to people around the world, you can become an amateur radio operator and do just that. In addition to just talking, you can experiment with two-way communications in television (SSTV) and computers (packet radio).

For some people, the subversive or the playful rulebreakers are more fun to check out. These radio hackers are broken into two groups: *clandestines*, who are the radio arms of political organizations that intend to overthrow their governments, and *pirates*, who are hobby broadcasters who might have a mission or might only be enjoying themselves for an evening.

Shortwave radio offers plenty of variety, and with today's miniature receivers, you can take your radio anywhere and listen to almost anything. Unlike the Internet (the information toll road), it's all free.

So, stop surfing the Internet for a minute and catch a ride on the shortwaves, the original information highway.

If you have any feedback concerning this book, either positive or negative, please feel free to drop me a line. Because of the large influx of mail, I can't promise a response for all letters, but I do read all correspondence, and I will try to respond.

Andrew Yoder
P.O. Box 109
Blue Ridge Summit, PA 17214
Internet e-mail: ayoder@delphi.com

Shortwave basics —an introduction

From reading some of the magazines and books that are currently available, you might get the idea that everyone in the world thoroughly understands technical shortwave applications: propagation, antenna construction, and antenna theory. But for some reason, the subject of shortwave radio has been held in the technical underground for decades. It's just been a real mystery to most people. However, with new publicity, low prices, availability of equipment, and benefits of shortwave listening, many people are being introduced to shortwave radio.

Rather frequently, someone sees my portable radio and starts asking questions. One of the most common is "You mean that radio can pick up stations from around the world?" So I usually tune in the BBC or maybe Deutsche Welle and let the person listen a little. The next question is usually "You mean this is coming from England?!" (or Germany).

Although most of the people who work with shortwave on a regular basis take it for granted, the distances covered by shortwave radio stations are truly amazing. And it's not just the huge government stations that get out this well. Even small, portable ham stations, mini-powered private broadcasters, pirates, and clandestines can be heard around the world.

The methods by which shortwave radio signals travel from the transmitter are a bit different than those in the standard broadcast bands. And few people are familiar with these methods. As a result, it's important to at least have a background in shortwave propagation, broadcasting, and listening if you are going to build your own antennas (see Chapter 2). This way, you can understand why certain aspects (such as antenna length) are important.

Propagation

Shortwave reception is an unpredictable, very complicated beast. The unpredictability is certainly one of the factors that attracted me into the hobby. It's exciting when you get a chance to hear stations that normally aren't audible in your region of the world.

Radio stations produce signals that take several different routes to reach their listeners. The two main types (although others do exist) are the ground wave and the sky wave. The *ground wave* is the radio wave that travels along the ground from the antenna tower. A good example of a ground wave is the signal that you can receive from a local AM broadcast station. The signal is generally solid and fade-free. Actually, several different types of radio waves are considered ground waves, but this is a simplified discussion of radio propagation. The *sky wave*, on the other hand, is a component of the radio signal that shoots into the sky at a higher angle and is bent back by the *ionosphere* (covered in greater detail in the following paragraphs). The sky wave often can be heard several thousands of miles away, depending on various conditions. Even strong sky wave signals often quickly fade down in strength, then return to the previous signal quality. This fading runs in cycles; depending on how fast or slow the fades occur, it could be either virtually unnoticeable or it could be quite bothersome.

Long-distance radio reception is possible because of one of the layers of our atmosphere that is known as the ionosphere. The ionosphere is many miles above the earth, where the air is "thin"—containing few molecules. Here, the ionosphere is bombarded by x-rays, ultraviolet rays, and other forms of high-frequency radiation. Energy from the sun ionizes this layer by stripping electrons from the atoms in the ionosphere. The ionized layers of the atmosphere make long-distance radio communications possible.

The ionosphere can, in turn, be divided into several layers, but the D, E, and F layers are the ones that affect radio propagation. The D layer is the one that is closest to the surface of

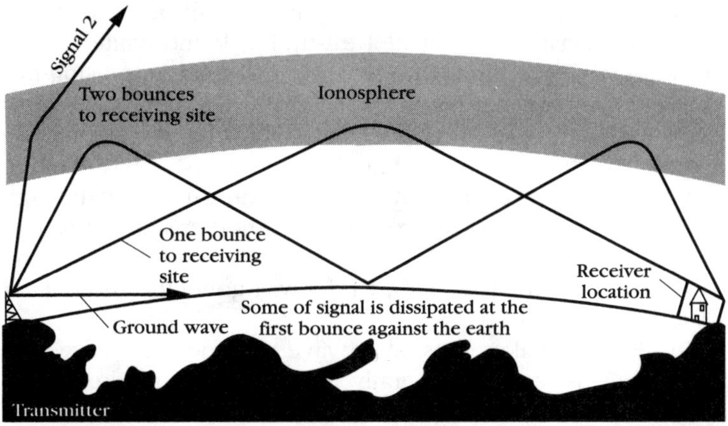

A simplified look at how shortwave radio signals can be received with help from the ionosphere.

the earth. The existence and strength of this layer depends on and is proportional to the position of the sun in the sky. As a result, the D layer gradually grows in strength in the morning, is strongest at midday, and gradually decreases until it disappears by nightfall. Also, the D layer is generally stronger in the summer than in the winter. On the lower frequencies (below about 10 MHz), the D layer absorbs any signals that are transmitted into it. For a great example of this effect, tune across the AM (medium wave) broadcast band at midday and then do the same at night. At midday, you can hear local stations and maybe a few powerful cross-state stations. But with nightfall, the band comes alive, and stations from across the country are audible. You have just witnessed the effect of the D layer.

The E layer is much like the D layer, except that it is a bit higher, and it has a few redeeming characteristics. The E layer absorbs most radio signals, but it sometimes refracts shortwave signals—especially those above about 14 MHz. So, during the summer months, the higher frequency amateur bands are particularly "hot" during the daytime when the lower frequency bands are being eaten alive by the D and E layers. One of the most interesting characteristics of the E layer is called sporadic E skip. During the sporadic E skip, some higher frequencies in the lower VHF region, which don't normally skip, suddenly skip for very long distances. Sporadic E skip is somewhat common, but you have to be at the right place at the right time for a good opening. One time I lived in a small apartment and watched television on a small, beat-up black-and-white TV that my sister had given me. One severe disadvantage to using this TV was that my sister had broken off the whip antenna a few years earlier. It didn't receive that well, but we were close enough to Pittsburgh that a handful of stations would come in well anyway. One evening at sunset, channel 4 from Pittsburgh started to fade out and was replaced by another signal. In a few minutes, KHAS, from Hastings, Nebraska, was in loud and clear. About 10 minutes later, KHAS was gone and the regular stations were back.

The F layer makes most shortwave and medium-wave skip possible. This layer is generally about several hundred miles above the earth, and it remains ionized throughout much of the day and night. However, unlike the other layers (under most circumstances), the F layer refracts signals back to the earth. Because the F layer is so high above the earth, signals often skip over very great distances from the F layer—sometimes several thousand miles. As a result, it is easy to hear international broadcast stations with clear signals from around the world.

The ionosphere, skip conditions, and signal absorption have a number of peculiarities. One oddity is that frequencies above the top of the shortwave band normally cut straight through the ionosphere and travel into space, but the lower shortwave frequencies are often absorbed by the ionosphere. As a result, there are no perfect frequencies—the best frequencies depend on the intended audiences, the time of the year, condition of the ionosphere, etc. Another variable that dramatically affects listening is the skip distance. Long skip is not necessarily the most desirable condition. For example, it is common when listening to the 20-meter amateur band during the daytime hours to hear a station from a distant country, such as Australia, talking to someone in your state. The problem is that you will probably only hear the station in Australia, not the one in your state! Thus, if you are an amateur operator, you need to try a frequency on a lower-frequency band if you wanted to contact someone that close.

The same holds true for broadcast listening. A number of U.S. private stations broadcast on the so-called 41-meter band. This area is fair to good for only close communications during the day, but at night, the signals get out for thousands of miles. At night (at my location), these stations often drop to about half strength and they rapidly fade in and out. The problem is that I am too close to the transmitter for good reception at that time and most of the signal skips over me. These problems are virtually unheard of at AM (medium wave) and FM broadcast frequencies, so most people are stunned that they occur on the shortwave frequencies.

One factor makes propagation forecasting and theory very complicated. As you know, shortwave signals are severely affected by the ionosphere, which is (in turn) affected by sunlight. Daylight and nightfall do not occur at the same time throughout the world. This creates some bizarre skip conditions, to say the least. The best way to see what frequencies are coming in at a particular time of day in your area is to either join a shortwave-listening club (so that you can see logs of what other people are hearing) or read the propagation forecasts in some of the amateur radio magazines. These methods are of much more value than simply trying to figure it out yourself.

This section has been just a very quick, very superficial look at shortwave propagation—just something to help you become acquainted with how shortwave signals reach different parts of the world. As stated earlier, shortwave propagation is much more complicated than this, and I left out many important terms and characteristics on purpose. Too much information scares some people away—thinking that the hobby is just too complicated for them. Fact is, although many aspects of radio (including shortwave) and television are highly technical, you don't have to be an engineer to enjoy these mediums. For a more detailed, yet easy-to-understand, coverage of shortwave propagation, see *The Practical Antenna Handbook—2nd Edition* by Joseph J. Carr (TAB/McGraw-Hill, 1994).

1

Shortwave equipment

The shortwave radio bands (from 1700 to 30000 kHz) are filled with hundreds, if not thousands, of interesting signals at any given time. You can receive international broadcasts in English; talk from amateur radio and CB operators; two-way traffic from government, military, ships, and aircraft; ship-to-shore telephone calls; faxes from weather stations; and "underground" messages and broadcasts from spies, radio pirates, drug dealers, and paramilitary groups. This book delves into the vast amount of shortwave broadcasts, but the other topics are beyond its scope. For more information on these other facets of the hobby, see Appendix B.

Dedicated shortwave radios (typically called *communications receivers*) are almost as old as shortwave itself, and broadcasting information or entertainment has always been at its heart. In the 1930s, shortwave radio was the *hot* technology. Everyone was amazed that they could hear the music from England and the increasing tensions in the propaganda from Hitler, Stalin, and Mussolini.

Communications receivers aren't the only radios that have covered shortwave. In the pre-FM days of the 1930s and 1940s, almost every AM radio also included shortwave bands. I wasn't there, but my dad was. He used to listen to classical music via shortwave at night in his attic bedroom on a beautiful chest-high Zenith console radio. It's still there, beside the old bed in my grandma's attic.

Today, times have changed. The world is running at a much faster pace. Not only do we want everything NOW, but we are ready to move to get it. Not only are we lucky if we stay at the same job for five years, but even when we do settle down, we need to fly and drive across the country to be able to keep up with business and the busy lives of our friends and relatives.

The shortwave radio is a wonderful key to keep up with the world at large and to enjoy a few pieces of it as we speed along with our daily lives. But technology is also speeding along and not all shortwave radios are built for traveling. For example, I could strap my grandma's Zenith console radio on my back and go for a walk (along with a car battery and a power inverter). But after rescue workers pulled me out from under the veneered wood cabinet and pried the battery out of my still-clenched hands, I would be forced into hernia surgery. After I returned from the hospital, I'm sure that the Zenith would return to the attic.

Over the years, portable shortwave radios have been developed, but most of the pre-1980s models were luggables--somewhat heavy radios that you could grudgingly drag off and listen to outside. The faux leather-encased Zenith Transoceanic models are excellent example of luggables. Most of the Transoceanics weighed in at more than 30 pounds (13.5 kilograms) and were both finely crafted and considered quite portable in their prime, between the 1930s and the 1950s. The Transoceanics were well known for their high quality, solid build, and status appeal (they were used by everyone from royalty to movie stars). Even so, except for the nostalgia, you wouldn't want to take a Transoceanic on any trip.

Throughout the 1950s and 1960s, only a few good-quality portable shortwave radios were built. Amateur radio was becoming mobile, but these radios were combination transmitter/receivers--heavy (by today's standards), expensive, and covering only the amateur bands. Most of the shortwave portables from this time period hit the international market in the late 1960s and early 1970s when several Japanese companies introduced stand-up transistorized radios. Unfortunately, most of these radios were just AM/FM models with shortwave added, apparently as an afterthought. The pioneers of the truly portable shortwave receiver were Sony, Grundig, and Panasonic, all of which introduced dedicated portable shortwave receivers in the 1970s. This development created an avalanche of portables that started in the 1980s and has continued through the 1990s.

In essence, if you want to pick up a radio for traveling, you are limited to purchasing a radio that was manufactured within the past 15 years. Since that time, a number of small portable radios have been designed. Because of the technological advances over the years, increasingly better radios have been placed in smaller boxes. Even some of the tabletop radios can be operated from a 12-volt dc power supply, suggesting that the manufacturer expects you to take it on a camping trip. In fact, it seems that some day, all shortwave radios will be portable.

Portable radio features

The cost of a portable receiver ranges from about $40 to almost $600. Fortunately, the shortwave radio market is rather small and only a few dozen different models are available. But obviously the sizes, styles, quality, and features vary considerably. A good receiver isn't very important if you only want to catch the news on the Voice of America. If that's the case, you can probably get by with a very inexpensive receiver. Even if you just want the cheapest receiver to perform the easiest tasks, you should still check out the different features so that you can find a receiver that best suits your needs. The following paragraphs cover some of the qualities and features of portable shortwave radios that you should examine before choosing a receiver.

> ***Receiver checklist***
>
> - ❑ *Is the receiver within your price range?*
> - ❑ *Will it perform up to your requirements?*
> - ❑ *Does it cover enough frequencies (preferably 500 to 30000 kHz)?*
> - ❑ *Does it have SSB reception or a BFO?*
> - ❑ *Is it small enough to take everywhere?*
> - ❑ *Is it built well enough to handle some physical abuse?*
> - ❑ *Does it have wide and narrow filters?*
> - ❑ *Does it cover the FM or air bands?*
> - ❑ *Does the radio's audio sound good?*
> - ❑ *Does it have a digital frequency readout?*
> - ❑ *Is the tuning rate adequate?*

Sensitivity

Sensitivity is the capability of a receiver to pull in weak signals. If the receiver has poor sensitivity, weak stations would not be audible on this receiver (strong signals are still audible on a set with poor sensitivity). If sensitivity is a real need for you, look at the receiver specifications, which are often printed in the catalogs of the major shortwave mail-order companies. Sensitivity is measured in microvolts, which might seem like a complicated measurement. But there's an easy way to check the sensitivity--the lower the number, the better. For portable receivers, you can't get much better than the Drake SW8 and the Lowe HF-150, both of which have an SSB (single-sideband) sensitivity of 0.5 microvolts.

Selectivity

Selectivity is the capability of a receiver to choose between different radio signals. For example, a very strong signal might be on 6010 kHz and another strong signal might be on 6020 kHz. On a receiver with poor selectivity, both stations would interfere with each other. A receiver with excellent selectivity has no problem separating signals that are just a few kilohertz apart. The selectivity of a receiver is based on the filters it contains; some companies build replacement filters to improve the selectivity of better receivers. The Sony ICF-2010 has been the only portable receiver that's popular enough to warrant such special treatment with replacement filters.

Image rejection

Image rejection is the ability of a receiver to be exposed to very powerful signals without "overloading." An extremely potent signal, such as that from a local AM broadcast station, might swamp the front end of the receiver. As a result, you might hear "images" on frequencies other than the one that it broadcasts on. Because of the small size and emphasis on low cost, portable receivers traditionally have been notorious for poor image rejection. I have had some receivers that, when connected to a long-wire antenna, swamped on a local AM broadcast signal. I could hear that signal in the background wherever I tuned. Very annoying. However, the image-rejection problems have been corrected (or at least improved) in many of the newer portables.

SSB/BFO

The *single-sideband (SSB) mode* is a special type of transmission that is much more efficient than traditional AM broadcasting. However, the fidelity is drastically diminished. Because of the reduced fidelity, only a few shortwave broadcast stations operate in SSB, but nearly every amateur operator uses this mode. The other problem with SSB transmitting is that when you listen to an SSB station in the AM mode, it will sound like some cartoon ducks--except less intelligible. To hear the SSB stations, you need a receiver that either has the SSB mode or has a beat-frequency oscillator (BFO). Receivers with the SSB mode are easier to tune than those with just a BFO, so this should be a consideration.

Receiver feature: Drake SW8

Unquestionably, one of the best portable receivers on the market is the Drake SW8. Here are some of its features:

- *Covers 0.5–30 MHz, 87–108 MHz (FM band), and 118–137 MHz (air band)*
- *Receives AM/USB/LSB on 0.5–30 MHz and FM on 87–108 MHz*
- *Three built-in bandwidths (6.0, 4.0, 2.3 kHz)*
- *AM synchronous detector for improved quality of AM signals*
- *Built-in whip antenna*
- *Requires either D cells or 110-Vac power*
- *High dynamic range*
- *FM stereo for headphone operation*
- *Operating parameters displayed on the front timer*
- *Backlighting can be shut off to conserve battery life*

The SW8 is excellent for all types of shortwave reception: The audio quality is superb, the tuning is smooth and is displayed in 100-Hz steps, and images from strong signals are rejected--even if large antennas are connected. Not only is the SW8 very pleasant to listen to, but the tough-signal reception is so good that I was the first person in North America to report one station (among other good catches), thanks to the SW8.

The downsides of the high-quality, well-built Drake SW8 are size and cost. The SW8 is approximately the same size as a typical modern tabletop shortwave receiver and the weight is about 10 pounds, or 4.5 kilograms (lighter than most tabletop models, but much heavier than most portables). The price is about $600, which is about three times the price of a typical portable but is still anywhere from $350 to $600 cheaper than comparable tabletop receivers. However, the cost is not prohibitive for the serious listener--especially if you need to pick up and go from time to time.

Digital or analog readout

Most adventurous people like to explore by going somewhere new, but they don't like feeling lost wherever they are. For this reason, digital readout (which prints out the numbers on an LCD) is much more popular than analog tuning (which moves a pointer in front of a few rows of frequencies). Because of the small size of most analog portables, few numbers can be written in the dial space and you wind up guessing at your frequency (±10 or 15 kHz). Fortunately, with the low cost of components for digital tuning, very few analog shortwave receivers of any type are made anymore.

One step beyond digital tuning is *station name tuning*, which is available on several Sony portable receivers. With this system, you can tune to some of the popular frequencies around the shortwave bands and the name of the station automatically appears on the LCD readout. On the Sony ICF-SW77, 94 of these memories are preset at the factory; you can program up to 162 different stations into the memories. Station name tuning is an unnecessary extra for me because I always tune the radio with a few shortwave newsletters, magazines, or books handy.

Tuning rate

One of the problems with the digital receivers is that most of them don't tune continuously across the bands. Some of the small digital receivers only tune in 1- or 5-kHz steps. Neither of these types of receivers are useful for hearing amateur radio stations, and you might have some problems with hearing some broadcasters, such as the Voice of Greece on 7448 kHz or Radio Copan International on 15674.6 kHz.

For the SSB stations, you might even have some problems with the digital receivers that tune in 100-Hz blocks. If the amateur is transmitting between the 100-Hz steps of your receivers, such as on 7240.15 kHz, the signal will sound "ducky," no matter where you tune. It can be a real problem when listening to some of the hobby pirate broadcasters that transmit music in SSB. Some of the better receivers tune finer than the digital readout can differentiate. For example, 100-Hz readout is common on the better portables, but some of these tune in increments down to 10 or so Hz. As a result, you can tune in SSB stations much better on these receivers (such as the Drake SW8).

Unfortunately, few of the shortwave mail-order companies list the tuning rate in their catalogs. If you plan to spend a little more on your portable and want to have an easier time tuning in the SSB stations, be sure to call the information line and ask the salesperson.

Size and weight

This book is about shortwave listening while traveling, so of course, size and weight are of great importance. Even though some of my old 60+-pound radios are excellent, they just don't cut it for traveling; they're difficult just to clean around, let alone carry. All of the "portable" shortwave receivers are at least relatively portable, but otherwise the size varies greatly. With the largest of the portables, the Drake SW8, you feel more like you're hauling a VCR with a handle than a portable radio. The SW8 is basically a tabletop radio that has been built tough for traveling. I wouldn't use it for business trips, but it's great for taking to a remote site for serious listening.

Most of the other high-grade portable receivers are close to the size of a large book. The Grundig Satellit 700, for example, is 12.25 inches × 7.25 inches × 3 inches (31.1 centimeters × 18.4 centimeters × 7.6 centimeters). The Satellit 700 is an excellent receiver. The only smaller (barely) comparable receiver is the

Sony ICF-2010. The next general size bracket for portable radios is similar to that of a small paperback novel. A few in this bracket are the Sangean ATS-606, Grundig YB-400, Panasonic RFB-45, and Sony ICF-SW7600G. These radios are all near 7 × 4 × 2 inches (17.8 × 10.2 × 5.1 centimeters). One of the smallest portables around is the shirt-pocket Philips AE-3905, which is only 2.75 × 3.75 × 1.25 inches (7 × 9.5 × 3.2 centimeters).

Some people like to quote the old "size isn't everything" adage. Well, in regard to shortwave portables, size *is* everything. The smaller, the better. But because of the high cost of technology and the cutesy novelty of small electronics, you have to pay for the tiny size with both higher prices and lost performance. In order to put the quality of one of the large portables into a novel-sized package, the manufacturer must increase the price. Also, in spite of the companies' efforts, the performance won't be as good, either. To get what you want, you have to make sometimes painful tradeoffs: "This size is too big, but I don't have the money for this feature, so I'll settle for this . . ."

Receiver review: Grundig YB-400

The YB-400 is one of the highest-rated small portables because of the high performance that the user gets from such a small, solid, inexpensive box. Here are some of its features:

- *0.54-30 MHz AM and shortwave, and 87-108 MHz FM (stereo on headphones)*
- *40 memory presets*
- *Direct-entry keypad*
- *Multifunction LCD readout*
- *Dual clock/timer and alarm modes*
- *Selectable tuning steps (1- and 5-kHz steps)*
- *Telescopic antenna and external antenna jack*
- *Adjustable BFO for SSB reception*
- *1.51 pounds (0.68 kilograms)*
- *Low price*

The YB-400 is very sensitive for such a small receiver. In fact, I was able to hear several low-power SSB pirate stations on the YB-400 with just the whip antenna. The stations were drifty, yet it was easy to "track" the drift with the BFO control. Furthermore, I used the YB-400 with a longwire antenna, and it was not susceptible to overload-

ing and being splattered with images from other strong stations. And this radio is very small, sturdy, and perfect for mobile listening.

Of course, although the YB-400 is good for general listening, it isn't a "DX machine" like the Drake SW8. And because of the small package, only a tiny speaker with a small compartment is provided--the sound from the speaker is tinny. However, the cost is only one-third of the SW8, and it can travel to many places that the SW8 can never go. Overall, for the size and the price, the Grundig YB-400 is tough to beat.

Batteries

The number of batteries that a receiver requires is closely related to the size and weight of a receiver. You can assume that a large receiver requires more batteries than a tiny one. At the top end of the spectrum, the large Drake SW8 requires six D cells. On the other hand, the tiny Philips AE-3905 requires only two AAA batteries for operation. Most portables require a few AA or C batteries (or a combination of the two). Of course, the batteries can add quite a bit to the weight. Some light receivers, such as the Sangean ATS-803A, require a stack of batteries that seems to weigh more than the receiver!

Not only is the weight of the batteries important, but battery life is a key consideration with a portable receiver. If your batteries croaked during a safari or a mountain hiking trip, the radio would be useless until you could get back to "civilization." Like the tuning rate, battery life is one of those specifications that's tough to find information for. Part of the reason for this lack of information is that the life depends on the type of batteries used and how loud the volume is. Still, it's worth calling some of the shortwave mail-order distributors to check if they can give you some helpful hints.

Can I have a second opinion?

If you are about to shell out big bucks on a new receiver, you sure don't want to be stuck with a lemon. What if the ad lied and those knobs do nothing more than spin? What if the $200 portable is just a $20 radio with a salvaged surplus calculator keypad? What if you can pick up more stations on the fillings in your teeth than on this radio? Before you buy, check out the equipment reviews from the following:

- IBS. *IBS produces the annual* Passport to World Band Radio *and the in-depth* Radio Database International (RDI) *white pages technical radio reviews. These reviews are some of the best you can find. They include technical specifications, plenty of hands-on radio talk, a star rating scale, and a price-versus-performance symbol.*
- WRTH. *For decades, the* World Radio TV Handbook *has been* the *annual source of information on radio and TV broadcasting in the world. The* WRTH *receiver reviews are less "flashy" and witty than their counterparts at* Passport, *but they are also excellent and very thorough.*
- Magazines. Popular Communications *and* Monitoring Times *both feature regular receiver reviews. The amateur radio magazines* QST, 73 Amateur Radio, *and* CQ *also feature radio reviews. These reviews are typically written by different people, so you might not get a real sense of consistency. However,* Monitoring Times *does feature columnist Larry Magne, who reviews receivers for* Passport *and* RDI. *You can learn a lot by reading these sources, and you won't have to wait until the next radio annual is available.*

Where to buy a receiver

If you have looked over some of the books and magazines that cover shortwave radio, you might even know what receiver you want to buy. Now what? In the United States, chances are that the only local store that carries any type of shortwave radios is Radio Shack, which features a full line of portable receivers, most of which are made by Sangean. Some other primarily nonshortwave companies, such as Sony, are getting some of their receivers into such chain stores as Sears. So, with some searching around town (provided that you live in at least a small city) at the various electronics stores and stereo shops, you might locate about half of the available portable receivers.

Radio Shack

Like many portables, the Realistic DX-390 contains scanning capabilities and a number of presettable memories.

The problem is that you won't see many of the good receivers. Radio Shack receivers *are* available at every one of their branch stores. A few Sony, Panasonic, and Magnavox receivers are somewhat common because these companies all make consumer video and audio equipment. So, one of the special dealers might decide to pick up a shortwave radio or two, just to see if they'll sell.

But some equipment manufacturers specialize in shortwave equipment, and you won't find their receivers everywhere. Grundig is one of the largest manufacturers of shortwave radios in the world; although they specialize, some of their radios are available in retail outlets. Sangean radios are most commonly seen under the Radio Shack name. Drake is legendary for innovative, high-quality U.S.-made shortwave and satellite receivers, but their equipment is rarely available outside of specialty shortwave shops and mail-order distributors. Lowe's description is much the same as Drake, except that the company is based in England.

What are the leading names in shortwave receivers?

You might be flipping through a general-interest magazine and see an ad for a Platypusonic ZR-2000 receiver. It can do everything! Receive all of the shortwave stations in the world without an antenna. Make shortwave sound better than FM. Why, you can hear stations all the way from the mysterious land of Russia! And because it's made in China, they can offer it to you for only $79.95! Sound too good too be true? It probably is. There aren't many manufacturers of shortwave radios, so tracking the good companies is easy. Although the following companies might have made a few "dogs" in their time, they are all dependable and have been building shortwave radios for years.

- *AOR**
- *Drake*
- *Grundig*
- *Icom*
- *JRC**
- *Kenwood**
- *Lowe*
- *MFJ*
- *Panasonic*
- *Philips/Magnavox*
- *Radio Shack*
- *Sangean*
- *Sony*
- *Yaesu*

**Denotes companies that make tabletop shortwave receivers, but have yet to make portables. If they ever get into the portable market, the results should be impressive.*

Another drawback of picking up a receiver from a local store is that it will probably be close to the list price. The major shortwave mail-order companies sell so many radios that they can offer portables for anywhere from $10 to as much as $200 off the list price. The only exception to this rule is Radio Shack, which generally offers its radios at good prices and sometimes has excellent sales. Of course, because Radio Shacks only sell the company's own receivers, you have to be yearning for a Radio Shack receiver. If you are, be sure to check all of the local stores for special sales and especially for the closeouts. During the closeout sales, receivers are often sold for 50% or less of the original price!

If you're as cheap as I am, you might be on the lookout for a used portable receiver. But the prospects of finding one are slim. Used shortwave radios are almost never advertised in the classified sections of newspapers or "shopper" papers. I go to plenty of hamfests/computerfests, where all sorts of amateur radio equipment, computers, consumer electronics, and parts are sold. Not many portable receivers even turn up here, but if you have a hamfest/computerfest coming to your hometown, it's worth a look.

There's a slim chance that you might find the portable receiver that you're looking for through the classified ads in the back of shortwave magazines and newsletters or through the shortwave/amateur sections of computer systems (such as HAM-FORSALE on the Fidonet). As with anything, there's a chance that you could get ripped off by ordering a radio through the mail from an unknown individual. However, the active shortwave and amateur radio communities are generally small, honest, and tight-knit. In one case, I saw where someone had ordered a radio from another hobbyist over one of the computer networks. The man who ordered the radio hadn't received the radio after a year, and he started complaining. He couldn't sue for his money back because the two lived in different countries. However, another man, an amateur operator (who didn't even know the buyer or the seller), was able track the "seller" who had run off with the money and somehow convinced him to return it. I have no idea what he said that made such an impression. This kind of helpfulness and accountability is very rare in most hobbies.

Other than these rather infrequent used offerings, your only hope for a used portable is to get a trade-in radio from one of the specialty shortwave mail-order companies. Most of these companies service their equipment and offer brief "return-for-any-reason" periods and 30 days parts and labor warranties. Because of their reputations and because most of these companies have been in business for decades, you can purchase a used receiver from them, worry-free.

Major shortwave mail-order companies

Looking for a particular shortwave radio that you can't find anywhere else? Want the best price for a receiver? Want a radio that's backed with a warranty? Or do you just want a good price on a used receiver? Try one of the shortwave mail-order stores. Although a handful (or two) exist, these are the largest:

ACE Communications
10707 E. 106th St.
Fishers, IN 46038
800 445 7717
800 448 1084

Atlantic Ham Radio
368 Wilson Ave.
Downsview, ON M3H 1S9
Canada
416 222 2506
416 631 0747

Barry Electronics
540 Broadway
New York, NY 10012
800 990 2929

Communications Electronics
P.O. Box 1045
Ann Arbor, MI 48106
313 996 8888
313 663 8888

Com-West
48 East 69th Ave.
Vancouver, BC V5X 4K6
Canada
604 321 3200
604 321 6560

C. Crane Co.
558 10th St.
Fortuna, CA 95540
Phone: 707 725 9000
Fax: 707 725 9060

Dubberley's on Davie
920 Davie St.
Vancouver, BC V6Z 1B8
Canada
Phone: 604 684 5981

Electronic Equipment Bank
323 Mill St. NE
Vienna, VA 22180
Phone: 800 368 3270
Fax: 703 938 6911

Gilfer Shortwave
52 Park Ave.
Park Ridge, NJ 07656
Phone: 201 391 7887
Fax: 201 391 7433

Grove Enterprises
P.O. Box 98
Brasstown, NC 28902
Phone: 800 438 8155

Lentini Communications
21 Garfield St.
Newington, CT 06111
Phone: 800 666 0908

Norham Radio Inc.
4373 Steeles Ave. W.
N. York, ON M3N 1V7
Canada
Phone: 416 667 1000
Fax: 416 667 9995

Tucker Electronics & Computers
1717 Reserve St.
Garland, TX 75042
Phone: 800 527 4642
Fax: 214 348 0367

Universal Radio
6830 Americana Pkwy.
Reynoldsburg, OH 43068
800 431 3939
614 866-2329

Accessories

An amateur radio operator often has a basement full of radios and a pile of little boxes that adjust levels, match levels, couple different boxes together, switch between antennas, etc. The way these people accessorize, Avon probably makes a killing off of them. Fortunately, portable shortwave radios are so simple that you don't really need to accessorize.

About the only accessories that would be useful for shortwave listening while traveling is a DSP or an analog noise reduction unit. *DSP (digital signal processing)* units are a broad category of computer-related electronics that receive an analog signal, change it into computer binary code, process that code with various algorithms, and change it back into an analog signal. DSPs are very useful for eliminating white noise, removing signals, limiting signals, and filtering audio.

DSPs have been used for decades, but they have just recently hit the consumer market because of the high cost of computer processing chips. In just the past few years, the new advancements in miniature computer chips have allowed more research into consumer products and parts that cost less. Now, DSP "black boxes" are becoming common in shortwave listening.

The DSPs might seem annoying to drag along on a trip, but if you have the space to take one, it can be worth it. Most motels and other public buildings are infested with radio interference from fluorescent lights, neon signs, and computers. Although the interference-reducing circuits in these DSPs can't entirely eliminate this type of interference, they can remove some of it. JPS Communications makes several units that are effective against this type of noise and also against *heterodynes*--high-pitched squeals that are the by-product of several overlapping radio signals.

But beware of most DSPs! The DSPs that are available in the shortwave market are typically from the amateur radio market. These DSPs are intended for better reception of narrow Morse code and SSB signals. As a result, the bandwidth is much too

The JPS NTR-1 DSP is excellent for removing heterodynes and wideband noise from shortwave broadcast signals.

narrow for standard AM audio. In most cases, I found that using a DSP filter made the audio of even a clear AM broadcast sound muffled, muddy, and unpleasant for listening. Presently, only JPS Communications builds some external DSP units specifically for shortwave listeners. If you are uncertain about which DSPs are good for general shortwave listening, ask a dealer which models are made for broadcast listening or ask if the DSP has bandwidths between 4 and 6 kHz wide.

The whys and whos of DSP

Why do you want DSP?

- *To reduce random noise*
- *To provide better filters than your receiver alone*
- *To remove heterodynes*

Who makes DSP units?

- *JPS Communications*
- *MFJ*
- *Timewave*
- *Ramsey*
- *Radio Shack*

Two innovative companies have introduced DSP into their shortwave radios--SGC and Watkins-Johnson. SGC offers several DSP functions that are integrated into an amateur/marine transceiver. Watkins-Johnson built an almost completely digital receiver that features programmable filters and 1-Hz digital readout. Although it seems doubtful that either of these companies will make a portable receiver any time soon, it will be interesting to see how long it will be before digital and DSP-integrated portables are on the market (thus reducing the need for DSP "black boxes").

DSPs aren't for every traveler. They generally cost upwards of $150, and they are about the size of a hardcover novel. So if space and cost are big issues with you, you're better off without one. But if you're looking for performance, a DSP might perfectly complement a Lowe HF-150 or HF-225, Grundig Satellit 700, Sony ICF-2010, or a Drake SW8.

Conclusion

One of the most difficult aspects of shortwave listening is simply choosing what equipment is best for you and then finding it. With just a little bit of research, such as calling for information and looking through catalogs, you can choose a receiver that performs well, is easy to use, and is convenient to carry. Now, you just need an antenna.

2

Signal fishing: Reeling in the big ones

If you want to hear anything on the airwaves, you've got to provide a way for that radio signal to get into your radio. If you want to fish, you have to have a fishing rod, enough fishing line, a hook, and some bait. If you want to listen to shortwave, you simply need an antenna. Using shortwave radio antennas is something like using a fishing rod, but you never have to worry about the signals "biting," casting into a knotted mass of reeds, or reeling in an angry snapping turtle.

One look at an engineering textbook on antenna design and construction, though, and you find a mass of equations printed on thousands of onion-skin pages. Yo! If you wanted be a rocket scientist, you would be at the Goddard Space Center. Shortwave antennas really aren't too difficult to install and use—especially when you're traveling somewhere. In spite of anything you might read in a communications magazine, newsletter, or book, the bottom line is getting a chance to listen to that radio. If you think that you must string up a 32-element log-periodic antenna at 60 feet (18 meters) anytime you want to listen to the shortwave, you'll never hear anything while you travel.

Take a gander at your portable receiver and you'll see that it has a retractable chrome whip antenna. It's got more segments than a millipede and it's long enough to disembowel a giraffe. You should be able to hear broadcasters from Tibet with this monster, right? Unfortunately, you are probably a better antenna than the mighty chrome behemoth. If you want to check it for yourself, wet your fingers and touch the antenna. Chances are that the signals will improve. You're a better antenna than the one on the portable, but don't bother mentioning it on your resume; you're still not a very good antenna, and jobs as antennas don't pay well.

So, if the whip antenna doesn't do its job, what then? It's easy to string up a quick antenna that greatly improves reception. The cost is almost nothing and the time spent on installing these antennas is also just above nil. Even with excellent portable radios, such as the Grundig Yacht Boy 400 and the Sony ICF-2010, using only the whip antenna results in a real loss of performance (about like coupling a $1000 stereo to surplus 89¢ speakers). These two shortwave radios perform on the same level as many tabletop radios that cost much more. Your radio is only as good as your antenna; even the most basic shortwave program listeners should do at least some limited experimentation with external shortwave antennas.

However, if you forget your antenna or don't have any wire handy (not everyone carries a spool of wire in their purse or suitcase), you can use the whip antenna with limited success. In a worst-case scenario, you shouldn't even worry about trying to string up another antenna. If stringing up a decent antenna requires too much time, then don't do it. The key here is to be able to hear some news or other interesting programming when you can, not to win a Nobel Prize for Shortwave Technology on the Road.

Popular fallacies about shortwave antennas

1. Indoor antennas are ineffective. *As long as the building is not metal or metal-structured, they should work fine.*
2. Loops in antennas cause large signal losses. *Loops might cause an antenna to snap, but they won't make a noticeable difference in the signal.*
3. The antenna must be as high as possible. *This is true for transmitting, but not for receiving,*

especially in the low shortwave frequencies and in the AM broadcast band.

4. Insulation or tarnish prevents the radio signal from entering the antenna. *Corrosion might ruin connections, but neither it nor tarnish ruin an antenna.*
5. Only specific-frequency antennas work well. *There is a case for this, but random-wire and longwire antennas are very effective, nonetheless.*
6. TV antennas are fine for shortwave reception. *They're OK in a pinch, but you'll notice a big difference between a TV antenna and any standard shortwave antenna.*

The important thing at this point is to get back to the basics. An antenna is necessary to receive signals. In the case of the portable radios with the built-in whip antennas, an external antenna should improve reception. Just how much the reception should improve depends on what it's worth to you to provide the auxiliary antenna. For example, when I go to my in-laws' house over the holidays, it's not worth it to me to install an elaborate antenna. I want to do some listening, but I don't want the time that I would spend setting up an antenna to cut into the holiday festivities. So, I settle on a low-class wire antenna that requires about 5 or 10 minutes of installation time. It's not great, but it's much better than just using the whip antenna. At other times, when I plan to do more listening and when I have more time, it's worth it to spend more time to install a better, more complicated antenna.

Likewise, a travelling businessperson would not be able to build a proper shortwave antenna and use it on a regular basis. The setup time, space, and weight required would be prohibitive. Besides that, chances are that he or she would be planning only to catch the news or programming from one of the large, easily heard shortwave stations. This being the case, an antenna that could raise the signal level from fair to good is all that would be necessary. That is exactly what this chapter is all about: simple, quick antennas that fill the minimum requirements for listening.

Connections

It's very simple to attach antennas to the new breed of portable radios. Some portable radios (such as the Grundig Yacht Boy 400) contain built-in jacks for plugging in external antennas for shortwave, some (such as the Drake SW8) contain built-in jacks for plugging in external antennas for the radio bands other than shortwave, but most (such as the Sony ICF-SW30) don't contain any jacks for external antennas.

For years (through the 1950s, '60s, and '70s), if you wanted a decent shortwave antenna, you had to build it yourself. The radio manuals didn't mention how to do it, and few publications were available for the shortwave listener. Then, the approach seemed to be that shortwave was just something to listen to while you prepared to be a radio amateur. In that time, you just read amateur radio books, experimented with building antennas and falling out of trees, and waited until that glorious day when you could graduate to being a ham operator. Amateur radio can be a great hobby for experimenters and people who like to talk. However, shortwave broadcast listening is also highly informative, and it's a lot of fun. By the late 1980s, some receiver companies recognized the real potential of shortwave listening, and they started selling complete shortwave kits: an easy-to-use portable shortwave receiver, a guide that explained what shortwave is, and a quick-install reel antenna.

The reel antenna is one of the handiest shortwave implements in years. Rather than forcing you to string an antenna or suffer through the weak signals from the whip antenna, the reel allows you to almost effortlessly put up a decent little antenna. And you can take it down and put it away in less than a minute.

Antenna stringing tips

1. Never string antennas where someone might trip or fall against them. *Avoid doing so even if you are by yourself in a motel room. You might forget about the antenna and hurt yourself or damage your radio.*
2. Avoid metal obstructions. *Get your antenna outside or against the windows of a metal-structured buildings (such as nearly all motels and large modern buildings) or the signals won't even reach your antenna.*

3. Ask permission to install large antennas. *Some motels have windows that open to a courtyard—perfect for a 100+-foot (30-meter) antenna. Ask for permission to install any such antenna or you will be held accountable if it causes any damage.*
4. Use the room furnishings to support your antenna. *When stringing wire around a room, loop it around pictures, over door frames, curtain rods, etc. This way, you won't damage the walls with nails or tape.*
5. Never use rocks to support wire antennas in trees. *Chances are good that you'll knot the rock in the tree and leave it dangling—a real health hazard. A better deadweight is a sock containing a few ounces of sand.*
6. Use a fine gauge of insulated wire for homemade antennas. *Large-gauge wire is heavy and bulky.*

If you don't have a reel antenna for traveling, but you want an antenna other than the whip (and you don't want to hire someone to act as your antenna), you can either buy or build an antenna. Reel antennas are inexpensive from shortwave dealers and Radio Shack, so buying tips are unnecessary. The following sections cover some simple antennas that you can build. For most of these antennas, you only need wire and a few easy-to-find parts. Soldering is an option in a few cases, but there are enough choices here that you won't even have to solder.

The basic approach for connecting an external antenna to a portable receiver is to strip about 2 inches (5.1 centimeters) of insulation off the end of a wire (almost anything from #8 to #30, though the most popular antenna gauges are from #14 to #22) and to wrap the stripped end around the end of the receiver's whip antenna. If you want to impress your friends, you can use the proper term for this type of connection and tell them that the antenna is *inductively coupled* to your radio. One problem with this method of antenna coupling is that it's easy to end up with an incomplete connection. It is really annoying to have to rewrap the end of your antenna around the radio's whip antenna every time you tune to another station, just to make sure that the connection is complete. A solution to the connection problem would be to pick up a pack of *clip leads* (also known as *test leads*) from an electronics store, such as Radio Shack. Each lead is a different color and has a shielded alligator clip on each end. These leads are really handy to have around to make emergency connections, to hook up several receivers to the same antenna, etc. It's a good idea to purchase a pack of these whether or not you want to use this as your main antenna-connection arrangement. With the clip leads, you only have to connect one alligator clip to the antenna wire and connect the other clip to the receiver's whip antenna. Easy!

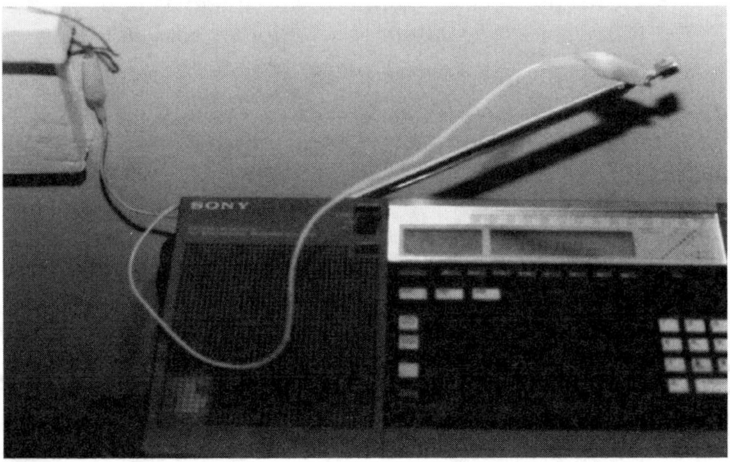

Clip leads connect the antenna to the whip antenna of a Sony ICF-2010 receiver.

To make the most permanent antenna connection for a portable radio, solder a connector on the end of the antenna. For radios with an external antenna jack, this would probably either be a ¼-inch (6.35-millimeter) RCA plug or a ⅛-inch (3.175 millimeter) plug (like the kind used on headphones for portable stereos). Make sure that you check what type of jack your radio has; otherwise, you might accidentally solder the wrong one onto your antenna. For radios without the external shortwave antenna jack, a standard alligator clip is the best connector to solder to the antenna. I would rather use clip leads to connect than plugs, because I'm prone to accidentally ripping off or breaking them. But some people prefer to solder the plugs.

The wire

The wire is the most basic form of antenna. The wire might not do everything well, but it does everything nonetheless. What I am calling the wire antenna is simply just a piece of wire, configured in whatever way. Unlike most standard antennas, the wire has no transmission line (used to carry signals from your antenna to your receiver), such as coaxial cable. This antenna functions as both the antenna and as the transmission line.

Like most forms of horizontal and vertical antennas, the wire can be almost any length. However, a longer, relatively straight version of the wire would generally be considered a *longwire antenna.* When does the wire become a longwire? I would hesitate to call a 5-foot (1.5-meter) piece of wire a "longwire" antenna. Some sources say that any antenna that's one wavelength long or longer is a longwire. Under this criteria, my 125-foot (37.5-meter) antenna would be considered a longwire above about 7500 kHz, and it would be considered a simple wire below about 7500 kHz. This sort of variable definition seems overly bulky and confusing for standard shortwave-listening purposes. Personally, I would say that anything over about 100 feet (30 meters) is a longwire antenna.

The receiving characteristics of a 100-foot (30-meter) longwire and a 1000-foot (300-meter) longwire are quite different, so if you're writing a reception report to a station or describing your antenna to someone, it's best to state the length of it and the direction that it is pointing. Wire antennas of less than 100 feet (30 meters) in length are much less directional in the way that they receive, so the direction in which shorter longwires are pointing is not as important for others to know.

As I mentioned, the receiving characteristics of a long longwire antenna are much different than those of a shorter wire antenna. The longer it is, the better it will receive from each end of the antenna. Likewise, the longer it is, the less signal can be picked up from off the sides of the antenna. This phenomenon occurs according to the proportion of antenna length to the wavelength (frequency) of the station that you're listening to. An extremely long longwire antenna might be a most incredible antenna for someone who only listens to a few small segments of the world.

For example, one weekend some friends and I got together to listen for low-power European shortwave pirate stations. We strung a longwire antenna that was somewhere between 1200 and 2000 feet (360 and 600 meters) in length (according to the campground map, the antenna was approximately 1700 feet—510 meters—long). The antenna was pointed to the northeast, right at Europe. As a result of our efforts, we heard several European pirate broadcasters with low power (only about 100 watts) and many North American pirates, including one low-power AM pirate from New York City. In the case of the New York City AM pirate, it was booming in with excellent fidelity. I decided to try hearing the station with about 30 feet (9 meters) of wire a few hundred feet away. Nothing. The massive longwire made the difference between hearing nothing and hearing an excellent signal. On the other hand, we found that many of the North American stations had better signals with a simple inverted V antenna because it was much less directional. As you can see, each antenna type has its advantages and disadvantages. It's all a matter of choice and taste.

One of the few drawbacks with today's crop of miniature portable shortwave sets is that the tiny electronics contained within the case are susceptible to electric currents—even signals within the radio. I heard a handful of stories about one excellent portable. Because of the quality of the radio, many listeners DX with it. As a result, the longwire is a logical choice to connect to this radio for DXing. However, if the signal into the receiver is too strong, it can overload the front end of the receiver and POOF!, a semiconductor or two are rendered useless. You won't even see smoke, hear a blood-curdling explosion, or smell molten plastic. You'll only notice a significant drop in signal strength. Although the radio is repairable, the $50 (or so) bill per pop should be prohibitive. So, the warning is: *The longwire antenna could put your portable radio at risk. Be careful!* Just how long is too long? I haven't gone over 400 feet (120 meters) long with my portable receivers (except for with the Drake SW8, which is built to handle stronger signals). Maybe I'm conservative with my antennas, but I like my radios, and I like them to be working!

Thus far, the only version of the wire that I've covered in this section has been the outdoor longwire. The wire can also be used indoors and it can be configured into different shapes. These antennas, which must be easy to install and remove, are the key types for portable shortwave radios on the market today.

As I stated earlier in this chapter, you shouldn't worry about little details, such as antenna resonance or impedance, with these indoor makeshift antennas. The big concern is just to get a length of wire up to "grab" as much signal as possible from the air. Crude as that might seem, it sure beats not having an antenna at all!

I have had moderate success with these types of miniature indoor antennas. I enjoy listening to music, news, or sports while working, so they have been particularly useful. For this type of simple program listening, these antennas often made the difference between success or failure at being able to understand the BBC news over the sound of running water at the sink while I washed dishes. As you might expect, these indoor wires are poor DX antennas—I have had a few good catches while using them, but these were rare and more based on extraordinary propagational conditions than on good antenna performance.

The reel antenna

One of the most practical new developments in shortwave-listening antenna technology (or gadgetry) is the *reel antenna*, which I mentioned earlier. The reel antenna is basically a handy version of the indoor wire that is neat and easy to carry. The actual wire antenna is held inside a reel. The other end of the wire has a small clip (somewhat similar to the type used to clip on chip bags), which you can clip onto the end of a whip antenna.

I received a commercial version of a reel antenna from one of my friends. I love having this antenna because I can wind it in immediately after a brief program-listening session in the kitchen. Because the regular wires were sometimes a pain to install and wrap up, I would normally just leave them strung up. This was always a problem because the wire usually crossed the kitchen cabinets and every time someone would reach in for a plate or a glass, they would be in danger of setting off my booby trap. The reel antenna has ended nearly all of these problems for me.

The reel antenna has one flaw—it's very short and it works for only the most basic listening. I wanted to experiment with constructing longer reel antennas—antennas that could actually be used for light-to heavy-duty shortwave listening.

The first problem was finding an appropriate reel. I didn't really want to modify a commercial reel antenna because it seemed to defeat the ingenuity of the experiment. I thought about using a fishing reel (fishing reel antennas have appeared in the hobby press for several decades), but I felt that it would be too bulky and that it would probably unwind into a mass of knots while I was taking it on a trip. My best guess for an appropriate reel was a standard chalk line, like those used in house construction.

I don't know how many people have old chalk lines sitting in their basements and garages, waiting to be converted into portable antennas. However, I would guess that few people do. A new chalk line isn't terribly expensive. In fact, I picked mine up for about $4—much cheaper than the price of a commercially produced reel antenna.

The huge advantage of using a new chalk line is that it hasn't yet been filled with chalk dust. If you choose a used chalk line to convert into a reel antenna, you must rinse it out several times with water. Otherwise, every time you unwind the antenna, the wire will be covered with chalk dust and bits of dust will spill out inside the house. Using a dusty chalk line for a reel antenna might work well for a vacuum cleaner ad on TV, but the resulting mess sure isn't fun in real life.

To build my reel antenna, I first pulled all of the string out of the chalk line reel until the reel was completely unwound. Then, I removed the screws in the chalk line case and pulled it apart. The string was easy to get to on the plastic spool, so I just cut it off with a pair of scissors. Next, I got the wire that I was planning to use, ran it through the spool, and knotted it. I used standard #20 stranded hookup wire with rubber insulation. After winding about 50 feet of wire onto the reel, I cut off the wire, stripped about an inch of insulation off of the end and soldered on an alligator clip. Then, I screwed the case back together (the case can be screwed back together after you knot the wire inside the spool).

I often string up longwire antennas when I visit my parents over the winter holidays. Invariably, I end up with a massive knot in my wire about 75 feet from the house and I spend about an hour trying to untangle the wire with numb hands in 10°F (–12°C) weather. After building the reel antenna, a light bulb popped up over my head and I imagined how convenient it would be to have a longwire reel antenna.

For the longwire reel antenna, I used very light gauge (approximately #26) enamelled wire. I chose this type of wire because it has a very small diameter and quite a bit of it could

A reel antenna being assembled.

be held on a chalk line reel. Also, the insulation is very hard and slick and is less apt to jam up the reel.

I was surprised to find that I could fit well over 300 feet (90 meters) of wire on the reel without having any winding problems. But one day when the temperature was about 25°F (–4°C), I was putting up the long antenna and the wire pulled off of the spool inside of the chalkline reel and locked up. I broke the case while trying to straighten the mess and I unceremoniously finished off the rebellious antenna by smashing it with a big rock. On my next attempt at reel antenna construction, I used less wire. It stays within the contours of the spool and works great.

The procedure for building the longwire reel antenna is the same as that for building the standard reel, except that I didn't use an alligator clip on the end. I was afraid that the clip would break off because of the light gauge of the wire, so I tied the end of the wire around the old end of the chalk line (to prevent the antenna from getting wound inside of the case). Then, I moved down the wire about an inch and carefully scraped off the enamel insulation for about an inch. When the antenna went up, I connected an alligator clip lead to the place where the insulation was removed on the antenna and connected the other end to the whip antenna of the portable shortwave receiver.

The reel antennas perform on par with standard wire and longwire antennas of equal length. The whole benefit of using a reel antenna is merely in the ease of installing and dismantling. However, in my winter experiences with the second longwire reel, I could wind it up in about five to 15 minutes (in the latter case, it ran through trees and into the woods, was covered with snow and ice, and it was several hours after sunset). I was pleased. I find the prospects of spending more time keeping warm while listening and less time freezing in snowstorms rather exciting!

The self-sticking screen antenna

The wire antenna and the reel version of the wire are excellent portable antennas. However, both are totally ineffective inside a metal-structured building, such as a motel, trailer, or office building. In these cases, it would be best to have an antenna installed outside. I have one friend who worked in an office building and his shortwave reception was terrible because the steel structure shielded out radio signals. He stealthily ran a very fine gauge of enamelled wire from his desk out through a window in his office building, and along a crack in the building to the ground floor. Although the system worked well, someone eventually found the antenna and removed it.

The bottom line is that using an outdoor antenna from a public building could cause you problems with your job or you could wind up with a lawsuit (if, for example, someone trips as a result of an antenna that you installed at a motel).

The only good location for an indoor antenna in a steel-structured building is at a window, where only glass is between the antenna and the great outdoors. A 3-foot (0.9-meter) wire antenna is impracticably short at shortwave frequencies, so that rules out the standard wire and vertical antennas. Surprisingly, one of the best makeshift antennas for this awkward predicament is a piece of window screen.

Window screen antennas have occasionally appeared in hobby publications over the past decade or two. In these publications, the window screen was normally assumed to be in the window. However, it seems that in most of the cases where this type of antenna would be necessary, there is no window screen in the window. Also, with the "wonders of modern technology," most of the newer window screens are made out of plastic, not metal. Because plastic does not conduct electricity, plastic window screens can not be used as antennas.

Essential to the window screen antenna is (obviously) a piece of window screen. In the ratio of size and performance versus convenience, a best bet for the size of this antenna would be between 1 × 2 feet (30 × 60 centimeters) and 2 × 3 feet (60 × 90 centimeters).

Try the local hardware store or construction supply center for window screen. Most of these stores sell various grades of window screen on large rolls. Once again, make sure that you use metal screen, not plastic screen. Either cut the piece to the size that you want or have it cut to size by a store clerk.

Afterward, you can place duct tape around all of the sides, except for a ½-inch (1.3-centimeter) gap along one of the sides. The duct tape around the edges will prevent you from cutting yourself or other objects while using or transporting the antenna. The small gap in the duct tape is where an alligator clip lead should be placed to connect the antenna to the whip antenna of the receiver. You might instead choose to seal all of the sides with duct tape (not leaving a gap for an alligator clip) and solder a wire directly to the screen. I didn't choose this means of connecting the screen antenna to the receiver because the solder connection would be rather fragile. I can just imagine a solder joint getting crushed and breaking in my luggage.

The next step is to find some small suction cups from a craft store. Lately, there has been a proliferation of small suction cups that are used for hanging such things as sun catchers on house windows and Garfield dolls in car windows. Just fasten one of these suction cups on each corner of the antenna and it's ready to hang!

When I constructed my version of this antenna, I used an old piece of metal screen that I found behind the wall in our laundry room. The screen was old, dirty, and slightly corroded, so I wire-brushed the area where I wanted to connect the alligator clip. Cleaning this area of the screen ensures a good contact between the clip lead and the screen.

As you might expect, the screen antenna is no great performer when compared to a standard outdoor antenna. But it does work somewhat well, and it is one of the best makeshift antennas if you're stuck in a metal-structured building.

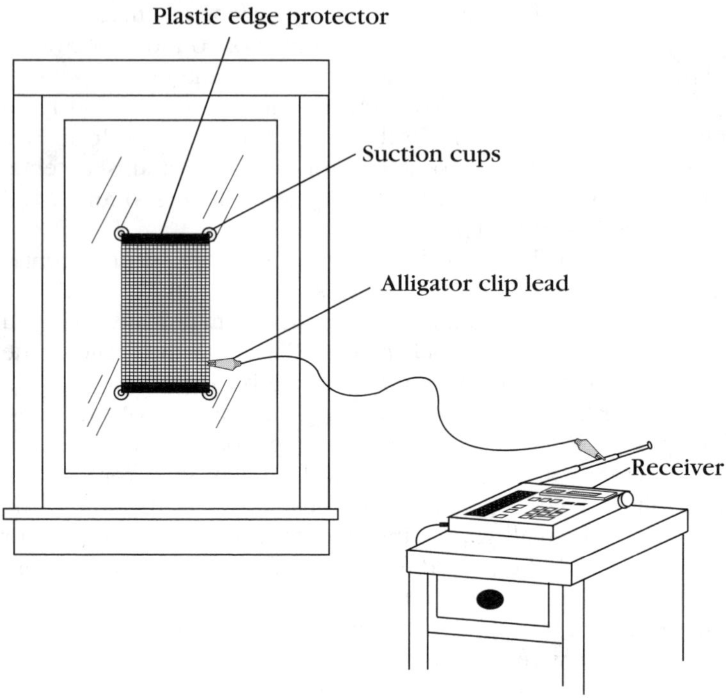

A window screen antenna.

Conclusion

With just a whip antenna, you won't "reel in the big ones." In fact, you'll probably be lucky if you get more than a few nibbles. But with nearly any sort of external antenna, your catches will greatly improve.

If these very simple antennas whetted your appetite for experimentation, plenty of other designs are available. For more information on other antennas for home or portable use, see my book *Build Your Own Shortwave Antennas* (TAB/McGraw-Hill, 1994). It contains these designs and a few dozen others to help you hear as much as possible—and have fun doing it.

3

Cruising the shortwaves: Radio in the car

If you thought that pulling out a portable shortwave radio and catching the "Slovak Kitchen" program was the ultimate, wait until you hear shortwave in the car! Car shortwave radios aren't typically portable, but they are mobile—they go anywhere that you can drive. If you don't usually go on business trips around the world and if you don't visit friends or relatives often, but your commute to work is long, check out car shortwave radios.

Auto audio shortwave equipment

For the most part, the only car shortwave radios that have been available in North America aren't typical shortwave sets. They are car stereos that also cover the shortwave bands. These models are poor for receiving rare shortwave stations, but the fidelity is excellent—far better than the shortwave portables.

Unfortunately, shortwave radio is virtually ignored by most everyone in North America, including auto audio manufacturers. Presently, the only company that manufactures *head units* (control units with a radio and either a CD player or a tape deck) with the shortwave feature and markets them in North America is the North American branch of Philips, the huge Dutch electronics firm. A number of Philips' head units contain the 48-meter shortwave broadcast band (just one small chunk of the shortwave spectrum), but this range would at least allow you to hear some of the shortwave giants (such as Radio Moscow, Radio Canada International, Radio Netherlands, Deutsche Welle, Voice of America, BBC, etc.) during the night-time hours. These 49-meter shortwave head units include the DCC-821 and DCC-811 DCC cassette decks, the DC-942 CD deck, the DC-722 cassette deck, and the DC-532 cassette deck.

Philips

The Philips DCC-811 head unit with coverage of the 48-meter shortwave band.

As far as shortwave listening goes, the real standout among the Philips line is (ironically) one of their lower-end units, the DC-777. This radio receives everything from 3170 to 21910 kHz (90 to 13 meters). Also, this unit makes shortwave, a medium that has never been known for high fidelity, sound bearable and sometimes even quite good—a great feat. The DC-777 has received a number of reviews in the shortwave hobby press and every one that I have seen has been very positive; it has probably received the most overwhelmingly positive reviews of any receiver that I have ever seen, most likely because of the lack of competition on this continent. The DC-777 can be a bit difficult to find from the regular auto audio outlets, but it is available from most of the major shortwave dealers, such as Gilfer, Universal Shortwave, Grove Electronics, and the Electronic Equipment Bank (see the mail-order section in Chapter 1 for more information).

About a year ago, I received word from Philips that the DC-777 has been discontinued and that "no successor has been planned for the time being." However, a few of these models are still available from Philips car stereo dealers and shortwave radio outlets. Hopefully, Philips will soon find a replacement for the DC-777!

The Philips DC-777 AM/FM/SW cassette radio with coverage of the shortwave bands from 3170 to 21910 kHz.

The 1991 Edition of the *World Radio TV Handbook* reviewed two other car stereos with shortwave reception that were available in Riyadh, Saudi Arabia. Although the correspondent mentioned that at least five different models were available, the only two that he specifically mentioned in the book were the Pioneer KE8700 and the National CQ800. Neither were rated nearly as highly as the DC-777, and it probably doesn't matter anyway. Chances are that in your part of the world, certainly here in North America, they aren't available. And, as fast as the different models of head units change, chances are that these particular models are no longer available.

However, some other car shortwave radios are sold in Europe. The 1994 Edition of the *World Radio TV Handbook* reviewed three car stereos with shortwave reception from Grundig, the well-known German shortwave manufacturer. These radios are all high-quality car stereos like the Philips DC-777, but unfortunately, the frequency coverage is limited to only 5950–6200, 7100–7300, 9400–9900, 11650–12100, and 15100–15600 kHz. Overall, the coverage isn't bad, but not nearly as good as the DC-777. Unfortunately, none of these Grundig radios have reached either the shortwave or the car stereo dealers in North America. As a result, I believe that these models are only being sold in Europe. So, if you want a shortwave/car stereo and can't find a DC-777, one of the Grundig models might be a good option.

If you live in parts of the world where shortwave plays an important role, such as in Europe, Africa, or the Middle East, you might find some other shortwave car stereo head units available. But if you live in a region, such as North America, where shortwave is mostly ignored, chances are that a used

Philips DC-777 will be one of your only options. Maybe some other company will see the practicality and usefulness of adding shortwave bands to an auto head unit and begin manufacturing them. If you don't live in Europe, either give up, go to Europe, get a trustworthy pen pal, find a used DC-777 . . . or get a shortwave converter.

Shortwave converters

Until recently, car audio technology has lagged far behind other electronics technologies. Most companies didn't even care if their radios sounded good, let alone if they covered nonstandard bands. The consumer should just be happy if they have an FM radio in the car, right?

This kind of mindset from the car audio companies caused many of the amateur and shortwave radio companies to experiment with car shortwave converters. These converters are little "black box" adapters that are mounted under the dash and are connected to the standard car radio. The converter changes the frequencies of the original car radio so that it receives the various shortwave broadcast bands.

Most of the old shortwave converters disappeared in the 1990s as the Philips DC-777 was introduced. The DC-777 was relatively inexpensive, the performance was solid, and the fidelity was excellent. The shortwave converters were rendered obsolete. Now that the DC-777 is no longer available in North America, the prominent American amateur radio accessory company, MFJ, introduced their MFJ-306 shortwave converter. Like other converters, this one operates from 12 volts (the car battery voltage), and it plugs into the standard car radio. The MFJ-306 covers most of the shortwave broadcast bands: 49, 31, 25, and 19 meters. Unfortunately, it doesn't cover the 21-meter band (13 MHz).

The pre-release information sheets from the manufacturer say that "the clarifier knob will let you tune in World Band stations lodged between standard AM stations on newer digital radios." Unlike some of the older converters, this one evidently allows both shortwave and AM (medium wave, or MW) signals to be received at the same time, making for disconcerting listening. And because there is no standard tuning knob on the converter, you have to push the tuning button on the car stereo (few analog car stereos exist anymore—nearly all are digital pushbutton models) and tune in the shortwave station with the clarifier knob of the converter.

The MFJ-306 shortwave converter, which covers the 49-, 31-, 25-, and 19-meter bands.

In spite of its limitations, the MFJ-306 is one of the only shortwave converters for car radios and stereos in the world. If you want more information from the manufacturer concerning the converter, contact MFJ at P.O. Box 494, Mississippi State, Mississippi 39762, USA.

Antennas

Often shortwave and medium wave (MW) listeners yearn to be away from the noise of the cities and tall buildings that obstruct the path of many of the rarer signals. A high peak or hilltop looks awfully inviting on those days when all that seems to come over the bands is garbled voices and ear-shattering static.

There might be times when you *really* want to listen to a particular program on a radio station that you cannot receive at home. Then you might want to drive somewhere, to a better location, to hear it. Unfortunately, given the monotonous state of radio programming that currently exists, this is doubtful; however, if you are either lucky enough to have a nearby community college or pirate broadcaster (but are still out of their signal's range), this might be the case. For example, in the movie *Pump Up the Volume*, people parked all over the local high school football field to listen to a pirate FM station because "that's where the signals are the strongest."

When setting up mobile antennas, remember that the land you are setting up on is not yours (unless you live on farm). Setting up could get you into a lot of trouble. Campgrounds and state and local government land are some of the best locations for mobile antennas. Just make sure that you follow any rules and regulations that apply to this land. Some locations, such as rest stops and picnic areas around major highways, don't seem to have any set regulations; you should have no problems installing low-profile, temporary antennas on this land. A few definite no-nos would be to string an antenna across the road or to run wire about neck height across the path leading to the bathroom. Just use common sense.

Private land is usually off limits. These days, many people are cautious about such things as lawsuits and drug deals. Installing an antenna on someone's land, then sitting in your car for a few hours, could make someone suspicious that you are up to no good. On the other hand, you could be on the land of someone who is simply paranoid. Trespassing is a good way to get shot or to land in jail for the evening. The food is bad and the reception is even worse, so don't do it!

Motor noise

When listening from a remote location in an automobile, you might decide to keep the engine running to maintain a charge on the battery or to provide heat during cold weather. Automobile engines sometimes provide a source of receiver noise that can range from slightly bothersome to completely overbearing. The ignition systems in cars can cause a severe buzzing interference in most receivers. The only real measures to remedy the situation are to either turn off the engine of the car, to use a diesel car (which doesn't have spark plugs), to properly ground the head unit, or to wire a noise-limiting capacitor in line with your equipment.

Vehicle antennas

Your parked car is like a portable shortwave listening station. You can obtain power from the cigarette lighter input jack on the dash or you can use a portable battery-powered receiver. You can also use a shortwave converter or a car stereo that covers the shortwave bands, such as the Philips DC-777.

Although this book is intended for casual listeners, the car listening trip is an interesting idea for active radio listeners who want to catch some rare stations without spending a fortune, losing vacation days from work, or neglecting friends or family for long periods of time. Even if your home location is good for shortwave listening, you can probably find a better place somewhere else. For example, I have a fair homesite for shortwave antennas, and my parents' home (which I occasionally visit) is excellent for listening. Still, I have had the urge to sneak out to the coast of New Jersey or Massachussets and point a huge longwire antenna at Europe. Radio waves lose more energy when traveling over land than they do traveling over water. So, if you are attempting to hear low-power shortwave or medium wave (AM broadcast band) radio stations from another continent in your car, you are better off listening from the coast nearest that continent.

Simple antennas for AM and shortwave hobbyists

Under most circumstances, the longwire antenna is the best choice for heavy-duty shortwave or AM (MW) mobile listening in a parked car or truck. All you need to do is remove the insulation from one end of the wire and attach it to the antenna terminal of the receiver or wrap it around the end of the whip antenna on the portable receiver. Then, run the wire out the window and keep going until you go reach the end and need to find a support to tie it to.

Very light gauge wire, such as #26, is excellent to use as a simple longwire antenna. Wire with such a small diameter is easy to wind, can be wound into a very small ball, and is difficult to see. Another advantage is that you can run the wire out through your car window and it will still close. Otherwise, you would need to pull out the head unit and build an antenna to specifically plug in for this application. This little extra edge is much more important when it is 15°F (–9.5°C) and you would otherwise be freezing to death with a ½-inch (1.3-centimeter) window opening for the length of coaxial cable.

Very long longwires are directional off the ends of the antenna (in other words, the signals are best received from each end of the antenna). So you should aim the wire at the direction that signal is coming from. Shorter longwires are less directional, but are still fine for general listening excursions.

Car antenna extender

The automobile antenna can be used to feed the signal to the shortwave receiver. An automobile antenna extension provides a means of lengthening the short whip that is supplied with most cars and trucks. You can lengthen this element to provide good AM (MW) or shortwave reception. The concept behind this antenna is just to turn your whip antenna into a quick-and-dirty longwire. This concept is about as simple as any ideas out there, and, best of all, it works well.

The construction is quite simple. A length of #16 copper wire serves as the major part of the antenna element. Slip the far end through the eye of a small ceramic insulator, then wrap it and solder it at several points. Scrape all insulation from the section of wire to be soldered. Thread the other eye of the insulator with a short length of nylon clothesline rope and tie it. Fit the opposite end of the element wire with a heavy-duty alligator clip. A clip lead similar to that which is supplied with some battery chargers is ideal and it is strong enough to hang on under the strain of long antenna elements. This lead should be firmly attached to the alligator clip and soldered at several different points. If you choose a smaller gauge of wire, you should be able to, in turn, use a smaller alligator clip.

The antenna extender is now complete. With typical antenna element lengths, the entire system can be stored in areas as small as the glove compartment of most automobiles.

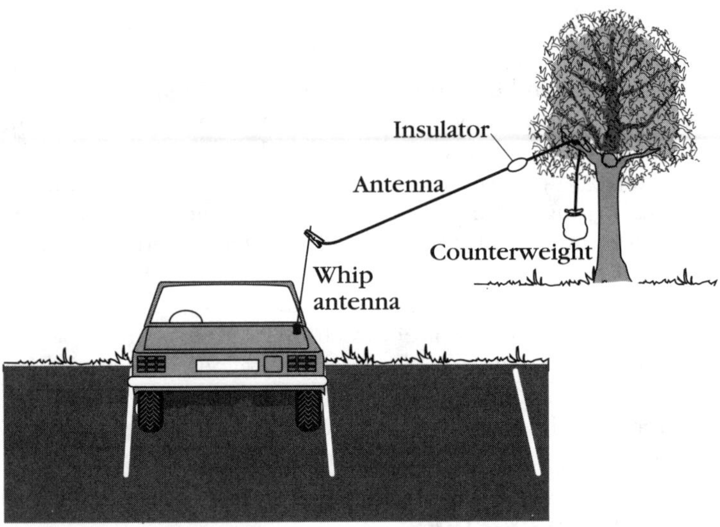

A simple car whip antenna extender.

Choose a likely support for the far end of the antenna element and attach it firmly. Carefully extend the element wire to the automobile antenna and make a firm connection at this point with the alligator clip. In addition to the usual caution about coming in contact with high-voltage lines, be certain that the *entire* system has been dismantled and stored before attempting to drive away from the portable operating location.

A practical antenna arrangement for active listeners

The longwire antennas and antenna extensions that are mentioned thus far in the chapter have been very simple, and no soldering or other complications have been necessary. The antennas and antenna systems in this following section are complicated, experimental, and (in one case) incomplete. You will probably want avoid these if you don't enjoy experimenting, but if you do, they can be a lot of fun!

You can do quite a bit to alter and boost the reception on both of the standard broadcast bands. The following system could be used in full or in part. I haven't tried it yet, and although I have seen very little information on these types of antenna systems (at least for this application), it should work well. After this book has been completed, I plan to be experimenting with this system. Maybe I will be able to report some helpful tips and fascinating information from these tests for the second edition of this book. Of course, if you also experiment with these (or other vehicular antenna systems), I will be interested to hear any information about how your systems worked out.

The antenna arrangement for this system is called a *phased array*. In a phased array, two vertical antennas are both fed with coaxial cable (similar to what is used for cable TV), which is connected to the receiver. Instead of being able to receive well from all directions (unidirectional), the phased array has a somewhat directional reception pattern. It receives best in two of the general directions, rather than all four. As a result, it receives better in directions that the array favors and poorer in directions that are nulled by the array. Thus, with an antenna system of this sort, you would be able to use it to receive some stations better.

Unlike what is often portrayed in the advertisements for the miracle television antennas, where you merely connect the junky "antenna" to the set and all of the signals suddenly become crystal clear, antennas don't really perform this way. Generally, there is a trade-off between directional and unidirectional performance. The more directional an antenna becomes, the less well it can receive in all directions. Directionality is generally terrible in a mobile system because as you turn on the road, the directions that the array receives best in keeps turning away from the location of the radio station's transmitter. In short, you constantly pull in, then null out the same station. If you are driving somewhere in Iowa, this might not be much of a problem, but here in Pennsylvania, where the roads curve more than they run straight, this system alone would be mostly useless.

Thus, you might need to correct these problems. One easy solution is to wire a three-way switch into the dashboard, near the head unit. In one position, only one antenna would be connected to the head unit. In the center position, both antennas would be connected together and form the phased array. In the last position, the other antenna would be connected to the head unit. With this switch configuration, you would be able to use the phased array whenever a weak station is coming from a direction that is within the "hot" direction of the array. If you then drove around a curve and the reception deteriorated, you could switch to one of the other antennas to improve the reception.

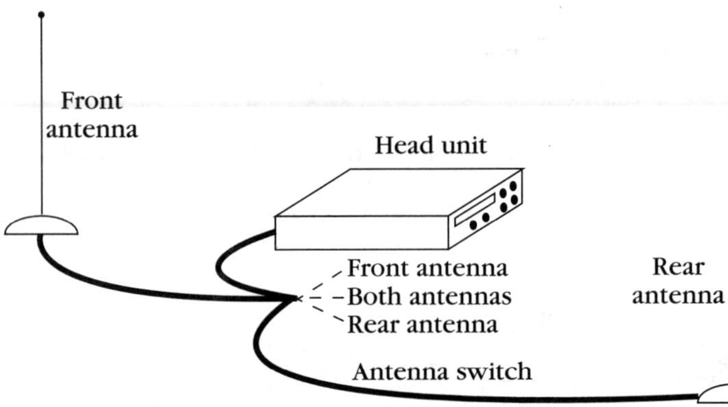

A switched antenna array.

Some of the questions that I have about this system deal with the transmission lines, impedances (a characteristic of antennas, coaxial cable, and radios), and the distance between the antennas. Normally, the antennas are arranged in a manner so that they are spaced a particular distance apart, depending on the wavelength of the antennas, impedances, etc. Under normal circumstances, if the antennas are placed closer or further apart than the predetermined spacing, the antenna receiving pattern will be skewed (the pattern will be less directional than if it was aligned properly). No equations for determining the spacing distance are included here because it will vary depending on the antenna, the locations for your antennas will be based on where you can attach them to the car rather than on mathematics, and because the distance is so short. As a result, you can find the most functional system by experimenting with different antenna locations until you find the spacing that works best.

Space-diversity reception

What would be great is if you could build an electronic gizmo to automatically switch between the different antenna configurations so that you wouldn't have to spend more time with your hands on the switch than on the steering wheel. Well, you can, but the system will be a bit different. Space-diversity reception is an arrangement where an electronic device receives signals from two (or more) different antenna sources and chooses the strongest signal for outputting. Rather than choosing between single antennas and the directional array, the space-diversity box chooses only between the two single antennas. I suppose that you could make it choose between two antennas that are in a directional array and another separate antenna. Then, the system would choose between the directional array signal and the unidirectional single-antenna signal. On the downside, you would be stuck with having three antennas perched on your car.

Unfortunately, you can't just pick up the coupling boxes for space-diversity systems anywhere. In fact, I don't know of any company that sells them. However, if you have an interest in electronics, then you might want to try building your own space-diversity coupling box.

Conclusion

For more information on choosing, installing, and maintaining car shortwave and stereo systems, see my book *Auto Audio* from TAB/McGraw-Hill (1995).

4

Listening on the road

You've got your portable shortwave radio. You've got some antenna plans. Now, you need to go somewhere and listen.

The hobby of listening to shortwave is one of its major misconceptions. You don't *have* to be a hobbyist who sits alone in a basement for hours, tuning the knobs and digging through the static. Some people learn a handful of words of Mandarin Chinese for the sole purpose of catching program details from the Chinese regional stations. Others send follow-up report after follow-up report to tiny Peruvian broadcasters with the hopes of receiving that treasured verification card someday. And the people who do these things regularly wake up at 5 a.m., drag out of bed, put on a pot of coffee, and scour the bands for rare broadcasters.

But shortwave listening can be fun and entertaining without any hard work. Armed with only a portable receiver, a recent shortwave guidebook or magazine, and an antenna, you can enjoy plenty of radio entertainment from around the world.

The key to enjoying shortwave is to make listening as easy and uncomplicated as possible. Overall, shortwave listening is a bit more difficult than flipping through the TV channels, but it doesn't have to be much tougher than that. About the only difference is that there are more channels and stations, and you have to press a few more (usually five instead of three) buttons to hit the right frequency.

The listening can be a bit tougher than the finding. Rather than the crystal-clear signals that you watch via cable TV, shortwave can be tinny, muffled, fading, or static-ridden. One friend of mine won't listen to shortwave because the noise gives him headaches; then again, he rarely ever listens to any type of radio, doesn't really like TV, and doesn't read newspapers. So, I suppose that the interference can be bothersome if you aren't interested in hearing news, music, or cultural events from around the world. Of course, not every signal that you tune to will be strong and clear, but many are.

If you find the static or weak signals to be particularly bothersome, just find a few of the strongest stations and listen to those. Remember that shortwave stations can be heard around the world and most of the major shortwave outlets beam their broadcasts to every continent.

For example, you might be listening to a particular station on a particular frequency in the United States. If you are traveling in Europe, you might hear the same transmission, but it might be weaker. Try another frequency that is beamed to Europe and see if the signal has improved. In other cases, a shortwave station might simulcast the same program on three different frequencies to North America. It's possible that one frequency will be inaudible, one will be fair in strength, and one will be quite strong. It could be a transmitter or antenna problem at the station, or it could simply be strange conditions or your particular location that affects your reception.

When to listen where

One common complaint about shortwave from beginning listeners is "I can't hear anything!" It might have something to do with the receiver, the location (such as inside a metal-structured building with no external antenna), or simply that the person is looking at the wrong place at the wrong time.

The wrong-place-at-the-wrong-time syndrome has to do with the locations of the shortwave stations, the frequencies that are being used, and "prime time" broadcasting:

1. *The radio conditions, known as propagation.* When listening to the shortwave, the main point that you need to know is that (in general) the frequencies above 10 MHz are best for daytime broadcasting and the frequencies below 10 MHz are best for nighttime broadcasting. The 31-meter band (9400–9900 kHz) is

useful for listening almost anytime. If you hear anything below 9400 kHz at midday, it's probably located less than 1000 miles away. In my location in Pennsylvania, USA, I can hear WEWN (Birmingham, Alabama) on 7425 kHz and CFRX (Toronto, Canada) on 6070 kHz at midday, for example.

The opposite occurs at the higher frequencies. The stations on 19 meters (15 MHz), which is excellent for listening at midday, disappear by about an hour or two after sunset. By three or four hours after sunset, you'll be lucky if you can hear more than two or three stations on the typically crowded 19-m band.

2. *Prime time.* Because most shortwave listeners go to school or have jobs, the stations target their programs for the largest audiences. As a result, if you are in North America, the prime time would run from about 2300 to 0600 UTC, depending on your location on the East or West Coast. There's still a fair amount of English programming to catch at other times (such as from the BBC, which seems to offer more programming than an entire cable network!), but it is typically beamed to other continents.

In summary, listen to the appropriate bands at the appropriate times and don't get disappointed if you're traveling in Europe and you can't hear CFVP (a tiny regional on 6005 kHz from Calgary, Canada) at 1300 UTC (midday).

What to pack

You've got the suitcase on your bed in the middle of teetering piles of clothes and you're trying to decide whether you have enough room to pack an extra pair of shoes or a can of shaving cream. With only 45 minutes before you hit the road, this is a lousy time to think about what you're going to eat tonight, let alone think about your shortwave radio.

It doesn't have to be tough. Most portable shortwave receivers are thin and wide and fit perfectly into a tight suitcase. If you're taking a road trip, keep it in your suitcase, unless you plan to catch the news somewhere en route. If you're going to an airport for a flight, take the radio as a carry-on. Some radios, such as the Grundig YB-400, even have a nice, padded vinyl carrying case to protect it.

Shortwave radio and flying

Because of the problems with airport security and international terrorism, airports are particularly sensitive about metal objects, especially electronics. Shortwave is an often misunderstood medium, and an unfamiliar-looking electronic gadget (your receiver) with lots of buttons might strike some fear into the security guards. Here are some tips to help you avoid problems at airports around the world:

1. Take your radio as a carry-on. *This way, you won't have to open your suitcase, you won't raise any suspicions, and the radio won't get damaged in the luggage bay.*
2. Keep working batteries in your radio. *If the security guards don't understand shortwave radio and are suspicious, you can turn on the radio for them and prove that it is a radio and not a remote-control device for a bomb.*
3. Don't make any "rough" modifications to your radio. *Perhaps you splintered the side of your radio in the excitement of a football (American soccer) game and you duct-taped it. Take that radio to an airport and everyone will be suspicious. Even if you tell everyone that it's a radio, they might believe you, but it will look like you've modified it for some other devious scheme.*
4. Take a copy of the owner's manual along on journey and keep it with the radio. *After you get used to your radio, you won't really need the manual. However, it could be helpful to show the manual to the airport security workers. Many manuals are written in several different languages, so even if the airport security guards don't understand English, they might know one of the others.*

Now that you've got your radio, you're set, right? Oops, you need an antenna if you will have a chance to listen in a somewhat stationary location, such as a motel or a friend or relative's house. Pull out the reel antenna or some wire. If you will only have a chance to catch some news or music while catching a milkshake at a fast-food restaurant, then stick with the built-in whip antenna of the portable.

You'll also need a power supply for the radio. Take the ac adapter if you are traveling anywhere in your home country. If you're traveling anywhere else in the world, leave the adapter at home. Chances are that the country you are traveling to has different power requirements or a different style of power plugs in use. With batteries as your only radio power supply, be sure to take along some extras if you are traveling to a remote area. If a large town or city is nearby, you should have no problem finding batteries in most Western countries.

Because not everyone enjoys overhearing "Let's Learn Chinese" from the Voice of Free China while you're sitting at Denny's eating scrambled eggs and toast, be courteous and take along a pair of headphones. Aside from having headphones with a plug that connects to your receiver (usually a ⅛-inch, or 3.175-millimeter, plug), make sure that the model you pick is sturdy. Over the years, I've broken a handful of the headphones that come packaged with personal stereos. Sometimes they've been smashed between books and other times they were caught in boxes. Either way, they break. Sangean packages up some perfect headphones for traveling with their shortwave radios. One model folds up and another has elements that fit inside your ears. I'm not particularly fond of the audio or of the "feel" of these headphones, but they are very easy to protect while traveling.

Information

Whenever you are navigating a car, boat, or plane in unfamiliar territory, you need a map to your destination. Otherwise, you'll get lost and you won't get where you wanted to go. Or maybe you'll get there, but you'll waste a lot of time along the way. Shortwave radio is much the same way; you need a guide or two to help you along the way, because there's just too much to be able to remember it all.

Of course, this book is a recommended companion for your trips. It's small, handy, friendly, and it doesn't snore. However, shortwave stations won't keep the same schedule until the next edition of the book comes out. Many shortwave frequencies are changed seasonally because the higher frequencies are traditionally better in the summer and the lower frequencies are better in the winter. Most stations are government-controlled and funding can be pumped in or cut at any time, such as the new transmitters and facilities at the Zimbabwe Broadcasting Corporation and the 1994 elimination of the Voice of America Bethany, Ohio, transmitter site. And all stations, whether they are government-controlled or not, are affected by political stability and violence, such as ELWA in Liberia and the Deutsche Welle relay station in Rwanda, which have been off the air during political upheavals.

With all of these changes, it's a great idea to subscribe to one of the shortwave magazines or monthly newsletters, such as *Monitoring Times*, *Popular Communications*, *Shortwave Magazine*, *NASWA* (North American ShortWave Association), and *ODXA* (Ontario DX Association). All of these magazines and newsletters cover shortwave broadcasting to one degree or another. Most of these feature late-breaking shortwave broadcast news, loggings from listeners, QSL columns, English-language broadcast schedules, pirate and clandestine broadcast news, receiver and book reviews, feature articles, and much more. To find the one that suits your particular taste, write for sample copies and pick the one(s) that you like the best.

Some of the shortwave magazines.

Monitoring Times
P.O. Box 98
300 S. Highway 64 W.
Brasstown, NC 28902
USA
704 837 9200
704 837 2216
BBS: 704 837 9200
$23.95 to U.S.
$34 to rest of world

NASWA—The Journal
45 Wildflower Rd.
Levittown, PA 19057
USA
$25 to North America
$32 to Asia, Africa, and Pacific
$29 to rest of world

ODXA
Box 161 Station A
Willowdale, ON M2N 5S8
Canada
905 853 3169
$2 or 4 IRCs for a sample

Popular Communications
76 N. Broadway
Hicksville, NY 11801
USA
800 853 9797
516 681 2922
516 681 2926
$21.50 to U.S.
$24 to Canada and Mexico
$26 to rest of world

Shortwave Magazine
Arrowsmith Ct.
Station Approach, Broadstone
Dorset BH18 8PW
UK
44 0202 659930
44 0202 659950

£22 to UK
£25 to Europe
$45 to USA
£27 to rest of world

After you have found a favorite magazine or newsletter, subscribe and carry the latest copy or two along on trips. It makes your listening more interesting, and it's a good read anyway. Good reading material is especially handy if you are traveling abroad and the water or the cooking doesn't agree with you.

Computers and listening

These days, everyone's using computers and exchanging information. Because of the level of technology involved and the number of people "surfing the 'net" and the BBSs, it's not surprising that a fairly large contingent of shortwave listeners are out there, lurking online. In fact, a few groups have even organized networks so that anyone with a computer, modem, and a phone line can keep abreast of the latest in shortwave information. Here are the three most-used shortwave computer nets:

- rec.radio.shortwave (Internet)
- SWECHO (Fidonet)
- Grove Enterprises (BBS)

The shortwave radio computer groups are excellent sources of immediate information. Listeners typically print new shortwave radio schedules, rare "catches," information about different programs, personal reviews of different shortwave radios, and ask and answer questions from other listeners. They also discuss all aspects of antennas and include discussions about the shortwave hobby in general. Some professional broadcasters and other important personalities (such as Andy Sennitt, editor of the *World Radio TV Handbook*) even participate in the discussions of shortwave radio. With top-notch listeners and shortwave professionals, you can get the latest information in just a few minutes.

This information is great because you can plug your computer in and download a few dozen pages of information (lists, tips, questions, stories, schedules, etc.) in just a few minutes, depending on the speed of your modem. I wouldn't advise buying a portable computer just to carry with a portable shortwave radio when you take a vacation, but the prices of laptops have dropped tremendously over the past few years. In fact, the laptop that I wrote this book with cost only $125. Regardless of the cost, probably more than half of the businesspeople who travel for their jobs now carry a laptop or notebook computer.

With a laptop, you only need to connect the modem port to a telephone jack, call the appropriate phone number and grab some information. The only problem is that at some locations, public phone jacks are tough to find. Only a few of the modernized telephone booths have phone jacks for computer links. Some motel rooms have standard modular phone jacks, so you can unplug the phone and plug in the modem. However, for the sake of security, some motels hardwire the telephones into the walls. In most cases, you can eventually find a location to get online, but it might be tough in rural regions (or outside North America).

There are a few differences between accessing the three shortwave computer outlets. *rec.radio.shortwave* is only available via the Usenet section of the Internet. In addition to the *rec.radio.shortwave* section, you can also send electronic mail to several of the large shortwave broadcasters. For these e-mail addresses, see Chapter 5.

To reach the Internet, you must have an Internet provider, such as one of many BBSs or a large commercial provider, such as America Online, Delphi, or CompuServe. The cost for a few hours per month of Internet is typically about $10 to $20. If you don't live near an Internet provider (my closest provider is about 40 miles away), you will also have to pay long-distance charges to your phone company. It's a bit expensive just for shortwave tips, but these days, there are so many advantages to tapping onto the Internet that chances are that you can find some other good reasons to log on.

If you are as cheap as I am, you might want to occasionally check out the BBSs, rather than pay a regular fee to an Internet provider. This way, you still have to pay the long-distance charges, but downloading 20 or 30 pages of information usually costs less than a buck. The SWECHO is a shortwave forum that is on the Fidonet. The Fidonet is, in turn, a number of different discussion groups that can be accessed from many computer BBSs around the world. You might have a local BBS or be visiting a location with a local Fidonet provider. If so, you might be able to check out the SWECHO for free.

The Grove Enterprises BBS is another major BBS that is worth checking into if you are interested in shortwave. Grove is a major name in communications—the company publishes *Monitoring*

Times and *Satellite Times* magazines and manufactures and distributes shortwave and scanner equipment. With the Grove BBS, you not only can find out the latest in shortwave information, but you can also get the inside track on their magazines and on the different radio products that are available.

QSLs and audience feedback

One of the big thrills of shortwave listening has always been writing to different stations from around the world. In the 1920s and 1930s, when radio technology was the cat's meow, almost everyone was excited—not just by what programming they could hear, but by what stations they could hear. In those days, which predated television, freeways, computers, fax machines, and regular passenger plane travel, radio conquered the seemingly impenetrable barrier of distance. For this great benefit, the populace wasn't just interested, they were *fascinated* with radio.

In the United States and Canada, radio magazines and newsletters sprung up everywhere. People built their own crystal radios and saved their much-more-valuable dollars for very expensive tube sets. Radio-listener clubs were started in most larger towns and most every town had a few technological geniuses—those who were amateurs and would build equipment for the local stations.

In the midst of all of this hype, some enthusiasts created contests—the winner of which had to hear as many different stations as possible. Because no one could verify if the listener had really heard all of the stations, the contest organizers went to the source, the radio stations, for help. To participate in such a contest, the listener would have to write down the date, time, and frequency that he or she heard the station and list some details concerning what was heard and send this report to the radio station. The radio station secretarial staff would read the report and compare it to the station's logbooks. If the report was accurate, they would send a special stamp or a logo card that stated that the listener had indeed heard the station and that the report was accurate.

Initially, the cards were called *listener response cards*, but they were later called *verification cards* or *QSL cards* (the latter being a Q-code from amateur radio). Today, most people have never heard of QSL cards, and few AM (medium wave) stations have the cards anymore, but nearly every shortwave

Andrew Yoder,Blue Ridge Summit,PA U.S.A. heard
Jolly Roger Radio-Ireland on 6.221KHZ on Jan.1st.1995
from 0933-0950. SINPO 13431.

This signal came from Waterford S.E. Ireland and the power
used was 120 watts,fed through a 70' di-pole antenna.
This was the only reception report from the uSA for this
broadcast,and was the first report from the uSA for 1995.

Verification information on the back of a QSL card. This one is personalized with special information about the specific transmitting equipment used for the test.

broadcaster still offers them. Indeed, most shortwave listeners have written for at least a few QSL cards in their lives.

Those listeners who collect QSLs seriously are much like the hypothetical listener at the beginning of this chapter. However, most of these listeners don't sit back and enjoy a few programs every week; that would be boring. These hobbyists are in it strictly for the challenge. I have even heard a few serious listeners say that they would sell their radios and quit listening to shortwave if they heard Tristan da Cunha, which was thought to be an impossible catch in North America. Silly!

Since you are reading this particular book, chances are that you might have no interest in QSLing the stations that you hear. However, I enjoy both types of listening (clear, interesting programs as well as rare "catches"), so maybe you will, too.

QSLs on the road

I have never written reception reports while I was traveling. For me, at least, it's too much of an inconvenience. If I'm traveling or visiting friends and relatives, I've got better things to do than write for QSL cards. However, if I have a chance to hear a new station, I copy down the details in my shortwave logbook.

RECEPTION REPORT

S Signal Strength	I Interference	O Overall Merit
5 Excellent	5 None	5 Excellent
4 Good	4 Slight	4 Good
3 Fair	3 Moderate	3 Fair
2 Poor	2 Severe	2 Poor
1 Barely audible	1 Extreme	1 Unusable

HCJB

English Service
BOX 17-17-691
Quito, Ecuador
Telephone
593-2-433-514 Ext. 441
Telex
22734 HCJB ED
Fax
593 2 447-263

NAME & ADDRESS: PLEASE PRINT

____

____

____

ANDEX MEMBER: Yes ☐ Number ______ No ☐

RECEIVER: ______________________________

ANTENNA: ______________________________

OCCUPATION: ____________________ AGE: ______

* Indicates information required for a QSL verification

DATE *	TIME (UTC) *	FREQUENCY *	RECEPTION			PROGRAM NAME*	PROGRAM DETAILS * & COMMENTS
			S	I	O		

To receive an AIR MAIL reply from HCJB, we request that you include one International Coupon (IRC) or unused postage stamps sufficient to mail a first class domestic letter (29 cents in the USA).

ADDITIONAL COMMENTS:

This special reception report form from HCJB makes writing to the station a bit easier.

A *logbook* is notebook or a three-ring binder in which you write information about what you hear. Include times, dates, frequencies, the strength and quality of the signal, and program information from the stations that you hear. When I get home, I copy the details over (unless I forget or procrastinate) into a somewhat standardized report that I keep on my computer. Also remember to mention the type of receiver and antenna that you are using so that the station can help gauge the strength and quality of their signal. Also, try to make some comments about the programming so that the stations can see how effective it is and, if possible and necessary, make adjustments in the programming.

If you are business-tripping and have a laptop computer along, you can always log the information right onto your computer. Some shortwave and amateur companies have even developed logging programs to make the computerized logging process very easy. On the other hand, most computers create so much interference that you might be better off not logging on a computer while listening. I use a trusty pen and a logbook, except for times when I catch a few programs at work. Then, I have to keep my computer on, so I'll just pop into another window and type in a few details while I work. Easy.

If you have a computer along, you can send reports or listener feedback to stations via the Internet. Because you can't send return postage via the Internet, this method isn't always acceptable. However, at this time, most of the stations that have Internet addresses are so large and well-funded that they will verify your report anyway. If they don't, I suppose they could e-mail you and let you know that they need return postage.

For more information on logbooks, writing reception reports to, and receiving QSLs from shortwave radio stations, see *The Complete Shortwave Listener's Handbook, Fourth Edition*, by Hank Bennett, David T. Hardy, and Andrew Yoder (TAB/McGraw-Hill, 1994).

5

International shortwave broadcasts in English

Just like on the Internet, you can become virtually lost if you scan through the shortwave bands without any information to help you find your way. This chapter lists countries, frequencies, times, addresses, telephone and fax numbers, and station features to give you a small taste of what's behind that schedule information.

How to use this chapter

Each listing features the following information (where available):

- *Country.* The country specifies the country of origin for the programs, not of the broadcast. Some stations have relay transmitters located around the world.
- *Station name.* Generally, the station name is listed in English, rather than that of the home service language.
- *Areas of the world that the station targets.* Some stations beam their programs to different parts of the world and others don't. The continent icons should quickly give you

a better idea about what you can hear, no matter where in the world you are:

- Africa 
- Asia 
- Australia 
- Europe 
- North America 
- South America 

- *Address.* The address is listed without the name of the country. If an address outside the country is used, that country's name is listed in capital letters.
- *Telephone and fax numbers.* Not all numbers were available, but many are listed. For more help on these numbers, try the *World Radio TV Handbook* (Billboard, annual) or the *Passport to World Band Radio* (IBS, annual).
- *Computer addresses.* Most of the large international shortwave broadcasters have Internet addresses, but some don't. Some stations, such as Radio Vlaanderen International, even have addresses on several different computer networks.
- *Prime frequencies and times.* The most common or regular times and frequencies for broadcasts in English are listed. The abbreviations that appear after some of the prime frequencies note to which continents those frequencies are broadcast:
  - **NA** is North America
  - **SA** is South America
  - **AF** is Africa
  - **EU** is Europe
  - **AS** is Asia
  - **AUS** is Australia
- *Notes concerning the transmissions.* The notes are brief comments about the stations operations—anything from the station being irregular to its future plans to its programming.

Albania

Radio Tirana

Radiotelevisione Shqiptar, Rruga Ismail Qemali, Tirana

Prime frequencies: 7160 kHz 0145–0230 UTC

7230, 9730 kHz 1800–1830 UTC

Notes: Once a powerful communist voice with a large English service, Radio Tirana has been drastically weakened by the changeover to democracy. An interesting development in light of the country's recent politics: The Christian broadcasting organization Trans World Radio announced in 1995 that it would be utilizing some of the old Radio Tirana transmitters.

Algeria

Radio Algiers International

21 Blvd. des Martyrs, Alger

Prime frequency: 11715 kHz 2000–2100 UTC

Notes: The station doesn't seem to put much emphasis on its English service—the times often change without notice between about 1800–2200 UTC, the frequencies sometimes change, and the broadcasts are often entirely absent.

Argentina

Radiodifusion Argentina al Exterior

CC 550, 1000 Buenos Aires

Prime frequencies: 11710 kHz 0200–0300 UTC **NA SA**

15345 kHz 1900-2000 UTC **NA EU**

Notes: Although Argentina is a large and important country in the Americas, shortwave apparently is not a top priority. The RAE is not regularly reported by listeners.

Armenia

Voice of Armenia

Alek Manukyan St. 5, 375025 Yerevan

Prime frequencies: 4810, 7480, 9675 kHz 1745–1800 UTC **EU**

11920, 11960 kHz 2300–2315 UTC **EU**

15170, 15270 kHz 0830-0900 UTC **NA**

Notes: This station was known as Radio Yerevan until 1995. Because of political instability in the region, listening to the Voice of Armenia is a search.

Australia

Radio Australia
GPO Box 428G, Melbourne 3001, Victoria
61 3 626 1800
61 3 626 1899
raust3@ozemail.com.au (Internet)

Prime frequencies: 5995 kHz 0630-0900 UTC
6020 kHz 0630–0900 UTC7240 kHz 0900-1130 UTC
7260 kHz 1800–2100 UTC
9580 kHz 0630–1200 UTC
9580, 9660 kHz 2100–0730 UTC
9615 kHz 1430–1800 UTC
11660 kHz 1430–2400 UTC
11695, 11800 kHz 1200–1630 UTC
11885 kHz 2200–0200 UTC
11880 kHz 1630–2100 UTC
13605, 15365, 15415 kHz 0030–0730 UTC
15245, 17880 kHz 0200–0730 UTC
17880 kHz 0200–0730 UTC
17795 kHz 0030–0630 and 2200–2400 UTC
21725 kHz 0800–1100 UTC

Also broadcasts in English on: 6060, 6080, 6135, 6150, 7205, 9510, 9610, 9645, 9770, 9850, 13745, 15170, 15240, 15510, 15530, 17715, and 17750 kHz.

Notes: Radio Australia is a powerhouse broadcaster in Asia and the South Pacific. In addition to a wide variety of programs available, Radio Australia has got to be *the* sports station on shortwave. The station dedicates many hours to broadcasting Australian sports (cricket, the Australian Football League, and the Australian Rugby League) every day.

Australia (Radio Australia)

Wednesday, December 20, 1939: five days before Christmas, the world is bracing itself for war. On the fringes of the Arctic, the Finns are locked in battle with the Russians. In South America, a freshly scuttled hulk of a Nazi battleship is settling on the bed of the River Plate after being trapped by British warships. Australia has just dispatched an advance party of servicemen to spearhead the war in the Middle East. And at a transmission station just out of Melbourne, a radio service is about to be born . . .

On the evening of December 20, 1939, the measured tones of the then Prime Minister Robert Menzies were heard inaugurating a permanent Australian overseas broadcasting service with the words: "The time has come to speak for ourselves." "Australia Calling," six years later to take the title "Radio Australia," was launched.

Fifty years later, Radio Australia celebrated its jubilee as the embodiment of the service which Menzies espoused. Acknowledged as having a major share of the shortwave audiences in many Asian and Pacific countries, it continues to speak for Australia and Australians in nine different languages, reaching an audience which numbers more than 50 million regular listeners. Every day, these people are informed and entertained by programs in English, Indonesian, Standard Chinese, Cantonese, Tok Pisin, French, Thai, Japanese, and Vietnamese.

— Radio Australia

Austria

Radio Austria International (ORF)
A-1136 Wien
431 0222 878 78 3636
431 0222 878 78 4404
kwp@rai.ping.at (Internet)
Prime frequencies: 5945 kHz 1930-2000 UTC
6015 kHz 0530-0600, 0630-0700 UTC
6155 kHz 0730-0800, 0930-1000, 1330-1400, 1930-2000 UTC
13730 kHz 0230-0300, 0730-0830, 0930-1000, 1130-1200, 1330-1400, 1930-2000 UTC

Also broadcasts in English on: 9655, 9870, 11870, 15410, 15450, and 17870 kHz.

Notes: Radio Austria International broadcasts news, special musical programs, and features about Austria's culture. The ORF's schedules proudly proclaim "Our editorial independence is guaranteed by law."

Radio Austria International. The English department, from left to right: Nigel Jones, Joanna King, Ann Dubsky, Merlin Koene, Patricia Maadi, Karen Engel, and David Ward.

Austria (ORF)

Shortwave is the classic and oldest means of international broadcasting communication. Austria counts as one of the pioneer countries in this field.

As early as 1929, experimental programs were being aired. However, at that time it was not possible to construct a high-power shortwave transmitter. Not until after the conclusion of the Austrian State Treaty in 1955 could a start be made on planning and preparing for a shortwave broadcasting center.

At Moosbrunn in Lower Austria, 25 km southeast of Vienna, transmissions started toward the end of 1959 using provisional equipment. On account of the various directional broadcasts sometimes in parallel in more than one language, more than 80 hours of programming are transmitted daily. Now, two 500-/300-kW transmitters and two 100-kW transmitters are used. Since 1989, our programs to North America are broadcast via satellite and relayed via Radio Canada International in Sackville. The English program can be received on the C-span audio network in the United States.

— ORF

Azerbaijan

Radio Dada Gorgud
370011 Baku
Prime frequency: 4957.5 kHz 1900-1930 UTC
Notes: Has also been noted at other times relaying the BBC World Service.

Bangladesh

Radio Bangladesh
P.O. Box 2204, Dhaka 1000
Prime frequencies: 7180 and 9650 kHz 1745–1845 UTC
9548 and 13615 kHz 1230–1300 UTC
Notes: Radio Bangladesh was operated as a clandestine station (known as *Swadhin Bangla Betar Kendra*) until the revolution

Radio Bangladesh.

succeeded in separating Bangladesh from Pakistan in 1971. Since then, Radio Bangladesh has slowly grown and modernized.

Belarus

Radio Minsk
ul. Kyrvonaya 4, 220807 Minsk
Prime frequencies: 5940, 7105, 7210, 7405 kHz 1945-2000 UTC
7150, 13650, 17655 kHz 0045-0100 UTC

Notes: Radio Minsk is rarely heard outside of Europe. With only two 15-minute English programs per day, there isn't much to hear anyway.

Belgium

Radio Vlaanderen
International (Brussels Calling)
P.O. Box 26, B-1000 Brussels
32 2 7328336
rvi@brtn.be (Internet)
2:292/878.5 (Fidonet)
32 3 8253613 (DXA BBS)
Prime frequencies: 5910 kHz 1800–1825, 2100–2125 UTC
6030, 13800 kHz 2330–2355 UTC
6035, 15545, 17595 kHz 0900–0925 UTC
9925 kHz 0630-0655 UTC
13670 kHz 1300–1325 UTC

Notes: *Brussels Calling* is the name for the English service of Radio Vlaanderen International. *Brussels Calling* features such programs as Belgium Today, Living in Belgium, The Arts, Press Review, and Music from Flanders.

Bhutan

Bhutan Broadcasting Service
P.O. Box 101, Thimphu
975 12072
975 23073
Prime frequency: 5025 kHz 1415–1500 UTC

Notes: Bhutan has been one of the fabled "rare radio countries" for radio listeners in North America and much of the world. Don't expect to hear them unless you are in Asia.

Botswana

Radio Botswana
Private Bag 0060, Gaberone
Prime frequencies: 4830, 7255 kHz 0300–0400 UTC

Notes: In addition to its national programming, Radio Botswana is well known for the barnyard animal sounds that it airs as an interval signal before and after its broadcasts.

Brazil

Radio Bras
CP 08840, CEP 70912-790, Brasilia DF
Prime frequencies: 15268 kHz 1900–2020 UTC **EU**
15445 kHz 1200–1320 UTC **NA**

Notes: Radio Bras is one of the only shortwave broadcasters from South America with English-language programming.

Bulgaria

Radio Bulgaria
4 Dragan Tsankor Blvd., 1040 Sofia
003592/87 10 60
Prime frequencies: 9700, 11720 kHz 0400–0500, 1900–2000, 2100–2200, 2300–2400 UTC **EU NA**
15635, 17625 kHz 1130–1230 UTC **AS**

Notes: Radio Bulgaria was formerly known as Radio Sofia. Unlike the broadcasters from countries that were part of the former USSR, Radio Bulgaria is regularly heard with strong signals.

RADIO SOFIA
BULGARIA

The resort of Pamporovo
Villégiature Pamporovo
Kurort Pamporovo

Radio Bulgaria International.

Cambodia

National Voice of Cambodia
Address is unreliable.
855 23 27319
Prime frequency: 11940 kHz 0000–0015, 1200–1215 UTC
Notes: Because of political instability within the country, the National Voice of Cambodia is not regularly heard.

Canada

CFRX
2 St. Clair Ave. W., Toronto, ON M4V 1L6
416 323 6830
Prime frequency: 6070 kHz 24 hrs
Notes: Relays local broadcasts from CFRB-1010 kHz in Toronto, Ontario.

CFRX from Canada.

CFVP
P.O. Box 2750, Stn. M, Calgary, AB T2P 4P8
403 240 5801
Prime frequency: 6030 kHz 24 hrs
Notes: Relays local broadcasts from CFCN-1060 kHz in Calgary, Alberta.

CHNX
P.O. Box 400, Halifax, NS B3J 2R2
902 422 5330
Prime frequency: 6130 kHz 24 hrs
Notes: Relays local broadcasts from CHNS-960 kHz in Halifax, Nova Scotia.

CKFX
2440 Ash St., Vancouver, BC V5Z 4J6
604 873 0877
Prime frequency: 6080 kHz 24 hrs
Notes: Relays local broadcasts from CKWX-1130 kHz from Vancouver, British Columbia. CKFX is currently off the air with transmitter problems, but when active, it uses a mere 10 watts of power!

CKZN
P.O. Box 12010, Stn. A, St. John's, NF
709 576 5099
Prime frequency: 6160 kHz 24 hrs
Notes: Relays national CBC broadcasts from CBN-640 kHz from St. John's, Newfoundland.

CKZU
P.O. Box 4600, Vancouver, BC V6B 4A2
604 662 6350
Prime frequency: 6160 kHz 24 hrs
Notes: Relays national CBC broadcasts from CBU-690 kHz from Vancouver, British Columbia.

Radio Canada International
P.O. Box 6000, Montreal, PQ H3C 3A8
514 284 0981
Prime frequencies: 5960 kHz 2300–0100 UTC **NA**
6120, 9535, 11725, 11845 kHz 0200–0330 **NA SA**
6150 kHz 1230–1357 UTC **AS**
11855, 17820 kHz 1300–1400 UTC **NA**
11945, 13650, 15140, 17820 kHz 2100–2230 UTC **AF EU**

Also broadcasts in English on: 6000, 7260, 9505, 9535, 9555, 9670, 9725, 9755, 11730, 11705, 11905, 11915, 11935, 11955, 13610, 15315, 15325

Notes: Even though Radio Canada International is one of the most popular shortwave stations, it has been threatened for years with elimination in the government cutbacks to the Canadian Broadcasting Corporation (CBC).

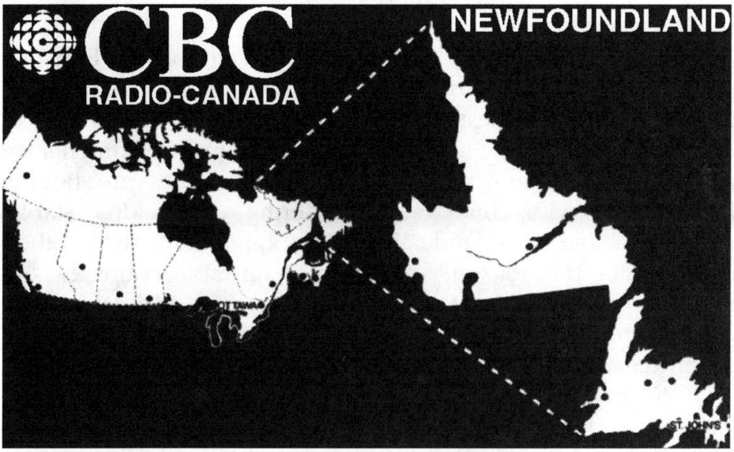

CBC Newfoundland.

China

Radio China International
No. 2, Fuxingmenwei St., Beijing 100866
6092274
8513135
8513174
8513175
8513176
Prime frequencies: 7405 kHz 1400–1600 and 1700–1800 UTC
9710 kHz 0000–0100 and 0300–0400 UTC
9730 kHz 0400–0500
11715 kHz 0000–0100 and 0300–0400 UTC
15440 kHz 1300–1400 and 2100–2300 UTC

Also broadcasts in English on: 3985, 6950, 6955, 7170, 9440, 9535, 9715, 9785, 9920, 11660, 11755, 15110, and 15130.

Notes: China Radio International is one of the last communist shortwave broadcasters in the world (along with Radio Pyongyang, Radio Havanna Cuba, and the Voice of Vietnam). CRI has one of the largest relay networks on shortwave, with outlets and/or relay agreements in Mali, Brazil, Canada, Spain, France, French Guiana, Switzerland, and Russia.

Clandestine broadcasters

Clandestine broadcasters are unofficial stations that transmit propaganda of some type, in order to overthrow or destabilize a government. Clandestines differ from political pirates in that clandestines are generally backed by foreign governments or revolutionary organizations, and political pirates are hobby broadcasters with controversial, differing views. Most stable countries have anti-clandestine radio agreements with other nations. For this reason, and because unstable countries are more apt to be plagued with clandestines, most of these broadcasters are in the world's hot spots operating in languages other than English.

In North America, the traditional target of clandestines is Cuba. Just a few of these clandestines have been La Voz del CID, La Voz del Alpha 66, Radio Caiman, and the infamous 1960s clandestine, Radio Swan/Radio Americas. The prime ground for clandestine listening is the Middle East, where literally dozens of stations have operated from such countries as Iraq, Iran, and Afghanistan. The only current English-language clandestine is Radio Free Bougainville, which has supported the rebels on Bougainville Island in Papua New Guinea, on 3880 kHz.

Clandestine stations typically air relatively simple propaganda, commentaries, music, and news. However, the programming heats up in time of war. Coded messages are sometimes broadcast and the clandestines are soon jammed by government transmitters. During World War II, particular songs, musical intervals, news items, and key words were used to signal members of the resistance within the Axis-controlled countries. To confound the effects of the propaganda, the Axis countered with fake versions of some of the clandestines. Everything became so complicated that some stations even claimed to be clandestines of the opposing side to portray their enemy in a poor light. One example was a German station with a former Scot as an announcer. The announcer was supposedly in the British Navy and he used excessively rough language and complained about "his" navy.

Although this type of clandestine activity is not active or audible in much of the English-speaking portions of the world, it could easily resume.

Prime frequencies: Most clandestines operate in the 7000- to 7100-kHz, 9900- to 10000-kHz, and 15000- to 15100-kHz areas throughout the day.

Cook Islands

Radio Cook Islands
P.O. Box 126, Avarua, Rarotonga
Prime frequency: 11760 kHz 1600–0500 UTC

Notes: Intersperses local programs with relays of news from Radio Australia and Radio New Zealand. Because of the low power, it is rarely heard outside the South Pacific.

Costa Rica

TIAWR (Adventist World Radio)
Ap. 1177, 4050 Alajuela

Prime frequencies: 13750 kHz 1100–1300, 1900–2000, 2300–2400 UTC.

Notes: One of the many branches of Adventist World Radio, TIAWR can be heard with a very strong signal across the Americas.

TIAWR, Adventist World Radio, from Costa Rica.

TIFC (Faro del Caribe)
Ap. 2710, 1000 San Jose
Prime frequencies: 5055 kHz 0300–0400 UTC

Notes: Faro del Caribe ("The Lighthouse of the Caribbean") is a Christian station that broadcasts almost entirely in Spanish but features a few English programs.

Radio For Peace International
P.O. Box 88, Santa Ana
SJO 577, P.O. Box 025216, Miami, FL 33102 USA
506 249 1821
506 249 1095
rfpicr@sol.racsa.co.cr (Internet)
Prime frequencies: 6200, 15050 kHz 1200–0000 UTC
7385 kHz 2100–0800 UTC
9400 kHz USB 24 hours

Notes: Radio For Peace International primarily broadcasts international left-wing programming from peace-oriented, human-rights, and women's-rights groups.

Croatia

Croatian Radio
Prisavlje 3, 4100 Zagreb
Prime frequencies: 5920, 7370, 9830, 13830 kHz 0703–0710, 0803–0810, 0903–0910, 1003–1010, 1303–1310, 2203–2210 UTC

Notes: Broadcasts regular programs only in Croatian but features news in English at the listed times.

Cuba

Radio Havanna Cuba
Ap. 6240, Havanna
radiohc@tinored.cu (Internet)
Prime frequencies: 6000, 9830 kHz 0100–0300 UTC **NA**
6000, 9820 kHz 0300–0500 UTC **NA**
9820 kHz 0500–0700 UTC **NA**
117050 kHz 2100–2200 UTC **EU**
6180 kHz 2200–2300 UTC **NA**

Notes: After losing support from the former Soviet Union, the hours of Radio Havanna Cuba have been cut back somewhat. Surprisingly, in spite of the strictness of the communist government, the "DXers Unlimited" is very popular among shortwave listeners, and the host, Arnie Coro, is somewhat of a shortwave celebrity.

Czech Republic

Radio Metropolis
Jeseniova 38, 130 00 Prague 3

Announced frequencies: 5905, 5940, 6200, 7250, 7305, 9455, and 9470 kHz.

Notes: Presently only testing with announcements in Czech, English, German, and Russian. It has been rumored that Radio Metropolis will not survive beyond the testing stage.

Radio Prague
Vinohradska 12, 120 99 Prague
cr@radio.anet.cz (Internet)

Prime frequencies: Radio Prague operates with 27-minute programs in various languages (including a number of English broadcasts) on 5930, 7345, 9420, 9505, 17485, and 21705 kHz. 5930 and 7345 kHz are most heavily used.

Notes: Some of the featured programs include Letter from Prague, Week in Politics, The Arts, Musical Feature, and Current Affairs.

Denmark

Radio Denmark
Rosenorns Alle 22, DK-1999 Frederikberg C
45 35 20 57 91
45 35 20 57 81
rdk.ek@login.dknet.dk (Internet)

Notes: Presently does not broadcast in English, but is considering offering English-language programming. All programs emanate via the transmitters of Radio Norway.

Ecuador

HCJB
Casilla 17-17-691, Quito
593 2 466 808
593 2 447 263
P.O. Box 39800, Colorado Springs, CO 80494 USA
719 590 9800
719 590 9801
english@mhs.hcjb.com.ec (Internet)
22734 HCJB ED (telex)

Prime frequencies: 6080 kHz 1430–2100 UTC
6135 kHz 0700–1130 UTC
6205 and 9420 kHz 0700–0830 UTC
9745 kHz 0030–0430 and 0500–1030 UTC
15115 kHz 1100–1600 UTC

Also broadcasts in English on: 12005, 15490, 17490 USB, and 21455 USB kHz.

Notes: HCJB is one of the most respected Christian broadcasters because of its projects (hospitals, hydroelectric plants, etc.) to help the people within Ecuador and also because of its interesting programming. In addition to regular Christian shows, the station also features the traditional music of Ecuador, El Mundo Futuro (about the future world of science and technology), and The Computer Corner.

Ecuador (HCJB)

"I can remember listening to Golos And *[Russian for HCJB's* Voice of the Andes*] ever since I was a child," says Roman Maximov. "However, I didn't commit my life to the Lord until 1987 when I was 14.* Golos And *has been the single most important source of spiritual food in my life."*

Growing up deep in Russia at Cheboksary, a city on the Volga River, 750 miles east of Moscow, Roman

Ecuador

knew what it meant to be a Christian. When Roman was a child, his parents would secretly listen to HCJB's Russian broadcasts on their shortwave radio. They knew that if the communist authorities learned about their faith in Christ, they risked losing their jobs and even being imprisoned. "Sometimes when we met in our little house church, the neighbors would throw rocks at the windows," Roman recalls.

Since the collapse of communism in Russia, Roman no longer has to hide his faith or conceal his short-wave radio while listening to HCJB. And the government no longer intercepts letters sent to the station. HCJB now receives more than 4000 letters from Russian-speaking listeners each year—10 times the number received before the increased freedoms entered Russia.

— HCJB

Egypt

Radio Cairo
P.O. Box 1000, Cairo
Prime frequencies: 9475 kHz 0200–0330 UTC **NA**
9900 kHz 2115–2245, 2300–0030 UTC **EU NA**
15255 kHz 1630–1830 UTC **AS**
15375 kHz 2030–2200 UTC **AS**
17595 kHz 1215–1330 UTC **AUS**

Notes: In spite of the good signals that Radio Cairo delivers to much of the world, the programming is often difficult to understand because of the rough-sounding audio.

Equatorial Guinea

Radio Africa
P.O. Box 851, Malabo
Prime frequency: 15190 kHz 2200–2300 UTC

Notes: Unlike the many national-service stations from Africa, Radio Africa is an evangelical Christian broadcaster that features plenty of programs that are produced in the United States.

Estonia

Estonia Radio
Gonsiori 21, EE-100 Tallinn
Prime frequencies: 5925 kHz 1620–1630, 2000–2030 UTC
Notes: Heard irregularly in Europe.

Ethiopia

Radio Ethiopia
P.O. Box 1020, Addis Ababa
7165, 9560 kHz 1600–1700 UTC
5990, 7110, 9705 1030–1100 UTC

Notes: Radio Ethiopia is not one of the more widely reported stations outside Africa. Still, the country often operates or is afflicted by clandestine stations, which are sought out by many. In 1995, the 9560-kHz transmitter was used for programs of Radio Amahoro, the Voice of Peace for Rwanda.

Finland

YLE Radio Finland
Box 78, 00024 Yleisradio
Box 462, Windsor, CT 06095 USA
800 221 9539 (US toll-free)
203 688 5540 (US)
3580 14805490 (Finland)
rfinland@yle.mailnet.fi (Internet)
Prime frequencies: 5990 kHz 2330–2400 UTC **AS**
6015, 9680 kHz 2330–2400 UTC **SA**
6120, 9560, 11755 kHz 0745–0800, 1930–2000 UTC **EU**
9635, 11755 kHz 0530–0600 UTC **AF AS**
11735, 15400 kHz 1130–1200 UTC **NA**
15330, 17800 kHz 0900–0930 UTC **AUS**
15400, 17740 kHz 1230–1300, 1330–1400 UTC **NA**

Notes: With contacts on several continents and an e-mail address, Radio Finland keeps in touch with its listeners—even if it is one of the smaller broadcasters in Europe.

France

Radio France International
BP 9516, F-75016 Paris Cedex 16

Prime frequencies: 6175, 11615, 12015, 15530 kHz 1600–1700 UTC

7110, 15405, 17560 kHz 1400–1500 UTC
9805, 11615, 13625, 15325 kHz 1200–1300 UTC
Also broadcasts in English on: 11700, 12015, 15155, 15195, 15365.
Notes: RFI is a major force on shortwave, but French-language programming takes top priority.

Georgia

Radio Georgia
ul. 68 M. Kostava, 380071 Tbilisi
Prime frequencies: 11805 kHz 0530–0600 UTC
11910 kHz 0700–0730, 1600–1630 UTC
Notes: Radio Georgia is not commonly heard outside the Middle East.

Germany

Deutsche Welle
50588 Cologne
P.O. Box 50641, Washington, DC 20091
49 221 345 2460
49 221 345 2461
49 221 389 4599
harald.schuetz.@dw.gmd.de (Internet)
solbach@dw.gmd.de (Internet)
www-dw.gmd.de (Internet program previews)
ftp-dw.gmd.de (Internet program previews)
Prime frequencies: 5960, 9515, 11705 kHz 0500–0550 UTC
6015, 9565 kHz 0700–0750 UTC
6040, 9640, 11865 kHz 0100–0150 UTC
6085, 9615, 11750 kHz 0300–0350 UTC
6170, 7225, 9875, 13690, 15595, 17800 kHz 1600–1650 UTC
6185, 7150, 9765, 11765 kHz 0400–0450 UTC
7115, 9670, 11765 kHz 2100–2150 UTC
7170, 9670, 11785, 13690 kHz, 1900–1950 UTC
7170, 9615 kHz 2000–2015 kHz
7235, 9690, 11705 kHz 2300–2350 UTC
7285, 9690, 12045 kHz 0200–0250 UTC
11915, 13790, 15185, 17820, 21680 kHz 0600–0650 UTC
12055, 15245, 17715, 21600 kHz 0900–0950 UTC
17860, 21600 kHz 1100–1150 UTC

Also broadcasts in English on: 9555, 11730, 11740, 11945, 11965, 15205, 17780, 17810, and 17875 kHz.

Notes: Deutsche Welle is regularly one of the most popular shortwave broadcasters across the globe. Like the BBC, the Deutsche Welle produces a *TV Guide*-type program listing every month, complete with descriptions, photographs, and behind-the-scenes features.

Deutsche Welle from Germany.

Ghana

Ghana Broadcasting Corporation
P.O. Box 1633, Accra
Prime frequencies: 3366, 4915 kHz 0525–2300 UTC

Notes: Broadcasts national programming in English, including news, music, editorials, and even the daily lottery numbers.

Greece

Voice of Greece
15432 Messoghion Ave., 432 Athens
01 6397 375
01 6396 762

Prime frequencies: 7448 kHz 0000–0350 UTC
9375 kHz 2200–2350 UTC
9935 kHz 0000–0350 UTC
15650 kHz 0800–0950 and 1300–1450 UTC

Also broadcasts in English on: 6260, 9425, 11595, 11645, and 17525 kHz.

Notes: The Voice of Greece is a relatively minor voice in the crowd of massive European shortwave outlets. The station is primarily intended to keep in touch with Greeks abroad.

Voice of Greece.

Guam

KSDA

Prime frequencies: 9370 kHz 1600–1700 UTC
9530 kHz 0900–1000 UTC
11980 kHz 2300–2400 UTC

Notes: One of the many branches of Adventist World Radio, the broadcasting arm of the Seventh-Day Adventists.

KTWR

1868 Halsey Dr., Agana, Guam 96922
671 477 2838

Prime frequencies: 11580 kHz 1500–1630 UTC **AS**
11830 kHz 0855–1000 UTC **AS AUS**
15200 kHz 0755–0915 UTC **AS AUS**

Notes: One of the branches of Trans World Radio, a Christian broadcasting organization.

KTWR from Guam.

Guatemala

TGNC (Radio Cultural)
Ap. 601, Guatemala City
Prime frequency: 3300 kHz 0300–0600 UTC

Notes: TGNC broadcasts throughout the day in Spanish and local languages, but it operates in English for several hours every evening.

Honduras

HRJA: Radio Copan International
8500 SW 8th St., Suite 252, Miami, FL 33144
305 267 1728
305 267 9253
71163, 11753 (CompuServe)
71163.1735@compuserve.com (Internet)

Frequency: 15675 kHz. Operates with various English- and Spanish-language programs on from 1100–0500 UTC.

Notes: HRJA is a small 1-kW station that operates as a sister station to WRMI in the United States. The main goal is to provide inexpensive transmitter time for individuals and low-budget broadcasters.

HRVC (La Voz del Evangelica)
Ap. 3252, Tegucigalpa
P.O. Box 828, Wheaton, IL 60187 USA
504 333 933
Prime frequency: 4820 kHz 0300-0500 UTC (Monday)
Notes: HRVC is operated by the Conservative Baptist Home Mission with many hours of Spanish-language programming throughout the day. However, the station does feature several hours in English every week.

Hungary

Radio Budapest
Brody Sandor u. 5-7, H-1800 Budapest
36 1 1138 8388
Prime frequencies: 3955, 5935, 7250, 9835 kHz 2100–2130 UTC **EU**
3955, 6140, 7130, 9835 kHz 1900–1930 UTC **EU**
6000, 9835, 11910 kHz 0230–0300 UTC **NA**
6000, 9835, 11910 kHz 0100–0130 UTC **NA**
Notes: Radio Budapest sticks to the well-tested, basic format of news and features.

India

All India Radio
P.O. Box 500, New Delhi 110001
Prime frequencies: 7140, 7412, 9910, 11670 kHz 1530–1545 UTC
7412, 9910, 9950, 11620, 11715, 15225 kHz 2045–2230 UTC
9705, 9950, 11745, 13750, 15145 kHz 2245–0045 UTC
9950, 11620, 11935, 13750, 15075 kHz 1745–1945 UTC
13732, 15120 kHz 1330–1500 UTC
15050, 15180, 17387, 17895 kHz 1000–1100 UTC
Notes: All India Radio provides an interesting look at one of the most populous countries in the world. Although All India Radio does not feature a huge English lineup, plenty of frequencies and programs can be heard worldwide.

Indonesia

Voice of Indonesia
P.O. Box 1157, Jakarta
Prime frequencies: 9675, 11752 kHz 0100–0200, 0800–0900, 2000–2100 UTC **AS AUS**
Notes: The Voice of Indonesia has announced that its transmitting facilities will be upgraded in the mid to late 1990s for better coverage.

Iran

Voice of the Islamic Republic of Iran
P.O. Box 15875/1575, Tehran
98 21 2041051
Prime frequencies: 7260, 9022, 9670 kHz 0030–0130 UTC **NA**
7260, 9022 kHz 1930–2030 UTC **AF AS**
11790, 11875 kHz 1530–1630 UTC **AS**
9670 kHz 2130–2230 UTC **AUS**
11745, 11790, 11875, 11930 kHz 1130–1230 UTC **AS**
Notes: Less Western-influenced than Radio Kuwait, the Voice of the Islamic Republic of Iran can provide listeners with a great step in understanding Middle Eastern politics and conservative Islam.

Iraq

Radio Iraq International
P.O. Box 8145, Baghdad
Prime frequencies: 13680 kHz 1230–1250 UTC **EU**
Notes: Schedule varies considerably, depending on the sanctions against the country and the conditions within. The station had a much larger schedule with better coverage before the Persian Gulf War.

Ireland

Jolly Roger Radio
P.O. Box 39, Waterford
Prime frequencies: most Saturdays and Sundays between 6219 and 6233 kHz at times between 0800 and 1500 UTC
Notes: Jolly Roger Radio is a regular hobby pirate that features country music and relays of other pirate stations from around the world.

Ireland (Jolly Roger Radio)

When I was small, I used to listen to a cheap transistor radio under my pillow. I was listening to a Trans World Radio broadcast one night and everything changed for me. When they closed down, Radio Geronimo opened up with rock music and they swore an awful lot, deliberate most of the time. I really got hooked on radio after this.

Back about 15 years ago, I was very friendly with the lads at ABC Radio. They had a small transmitter of about five watts. Shortly thereafter, they had the transmitter at my home. Because I was in the habit of listening to shortwave pirates on Sundays, it was no fun listening to ABC Radio all day (it's signal damaged listening across the 48-meter band). As a result, I started making taped programs as Riverside Radio International for five or six years for this transmitter until November 1988, when Jolly Roger Radio was born. We were the only Irish station on shortwave for years after the close down in 1988; all the others were too scared to come on.

Jolly Roger/Ireland

Our transmitter is homemade and is rated at 120 watts. We have been picked up in most European countries and we have also been heard in New

Zealand, Indonesia, Uruguay, Ecuador, USA, Canada, Russia, and Morocco.
—Joe Vincent, Jolly Roger Radio

Radio Dublin
P.O. Box 2077, 4 St. Vincent St. W., Dublin 8
Prime frequency: 6915.5 kHz, irregular

Notes: Radio Dublin is one of the oldest pirate stations in the world, with approximately 30 years of operations on AM, FM, and shortwave. Radio Dublin is sometimes tolerated by the government, but it has been raided and taken off the air many times over the years. Because the only shortwave operations from Ireland are via pirates, Radio Dublin is the unofficial voice of the country.

Israel

Kol Israel
P.O. Box 1082, Jerusalem 91010
972 2302 327
7465, 9435 kHz 0400–0415, 1900–1910 UTC **EU**
11605, 11685 kHz 1900–1910 UTC **AF EU**
15640, 15650, 17575 kHz 1000–1030 UTC **NA EU**

Notes: Kol Israel has been fighting off budget cutbacks throughout the early 1990s. Although the entire shortwave operation was threatened to be closed down, Kol Israel is still continues to broadcast.

Italy

Italian Radio Relay Service
P.O. Box 10980, I-20110 Milan
39 2 7063 8151
100020.1013@compuserve.com (Internet)
Prime frequency: 7125 kHz 24 hrs

Notes: The IRRS is another "transmitter for hire," as the name describes. In addition to relaying programs from large broadcasters, the IRRS also dedicates several hours every Sunday morning to airing programs from European pirate and hobby broadcasters, such as Radio Pamela, Southern Music Radio, International Music Radio, etc.

Radio Europe
c/o Play DX, Via Davanzati 8, I-20158 Milan
Prime frequency: 7294 kHz USB 0830–1200 UTC
Notes: Radio Europe is primarily a relay transmitter for European pirate broadcasters, such as Radio Marabu, Level 48, Radio Joystick, etc.

RAI (Italian Radio and Television Service)
Viale Mazzini 14, 00195 Rome
39 6 322 6070
Prime frequencies: 5990, 7275 kHz 0425–0440 UTC **AS EU**
7235, 9710, 11800 kHz 2025–2045 UTC **AS**
7275, 9575 kHz, 11905 kHz 1935–1955 UTC **EU**
9645, 11800 kHz 0050–0110 UTC **NA**
9710, 11800, 15330 kHz 2200–2225 UTC **AS**
Notes: Surprisingly, RAI (Radio Rome) is a minor broadcaster that broadcasts in English for only about 20 minutes a shot.

RAI from Italy.

Japan

Radio Japan
2-2-1 Jinnan, Shibuya-ku, Tokyo 150-01

Prime frequencies: 5960, 9680, 11840, 11910, 17845 kHz 0100–0200 UTC **AS NA**

5975, 6025, 7230, 9565, 11740, 11885, 15410, 17810 kHz 0500–0600 UTC **AS EU NA**

5975, 7230, 11740, 15270, 15335, 15410, 17810, 21610 kHz 0700–0800 UTC **AS AUS EU**

6035, 9560, 9580, 11800, 11875, 11925 kHz 2100–2200 UTC **AF AS EU**

6120, 9610, 15295 kHz 1100–1200 UTC **AS NA**

6150, 7140, 9535, 9580, 11850 kHz 1900–2000 UTC **AS AUS NA**

6150, 9535, 9580, 11930 kHz 1700–1800 UTC **AS NA**

6055, 6155, 99560, 9580 kHz 2300–2400 UTC **AS EU**

9535, 9750, 11705, 11840, 11915 kHz 1400–1500 UTC **AS NA**

9535, 9750, 11955, 15355 kHz 1500–1600 UTC **AF AS NA**

9680, 11840, 17810 kHz 0300–0400 UTC **AS**

5960, 11885, 11895, 15230 kHz 0300–0330 UTC **NA**

9610, 9750, 11815, 15190, 15270 kHz 0900–1000 UTC **AS AUS**

Also broadcasts in English on: 5965, 9660, 11705, 11860, 11865, 11895, and 11900 kHz.

Notes: Tune in to the voice of one of the most powerful economic forces in the world, Radio Japan. The station is heard well throughout the world via relay agreements and transmitter sites in England, Singapore, Canada, Sri Lanka, Gabon, and French Guiana.

Japan (Radio Aum Shinrinkyo)

You won't find a listing for Radio Aum Shinrinkyo anywhere in this book for several reasons: 1) It's no longer on the air; 2) It wasn't a real radio station—it was a program that purchased time on the Voice of Russia transmitters.

So, what was Radio Aum Shinrinkyo, and why is it featured here? In early 1995, terrorists attacked a Tokyo subway with deadly nerve gas. For weeks, this story and related subway gas attacks in Japan regularly topped the news around the world. After some investigation, the attack was alleged to have been committed by members of the Aum Shinrinkyo organization, which has been touted by many as a cult. The leader of the group, Shoko Asahara, has denied any involvement with the subway gassings, but Aum Shinrinkyo controlled plenty of the chemicals used to make the gas.

The voice of the group, Radio Aum Shinrinkyo, was heard across the world as the Voice of Russia allowed its relay transmitters to be rented in the early 1990s. According to Newsweek, *the group paid $800,000 per year to have its broadcasts regularly aired by the Russian Mayak shortwave transmitter site. Many shortwave listeners tuned in and quickly tuned out, quickly bored by the almost nonstop chants. With the investigation of the group, the Voice of Russia pulled its relays of Radio Aum Shinrinkyo in March 1995.*

With the Japanese government legally eliminating Aum Shinrinkyo in July 1995, it is doubtful that you would be able to contact the group through their Japanese contact numbers or address. But if the upcoming events and information from this former broadcaster intrigue you, try their American office:

8 E. 48 St., #2E
New York, NY 10017
USA
212 421 3687

Jordan

Radio Jordan
P.O. Box 1000, Amman
962 6 788 115
Prime frequency: 9560 kHz 1200–1300, 1500–1600 UTC **EU**
15170 kHz 1100–1200 UTC **EU**
15270 kHz 1200–1630 UTC **NA**

Notes: In years past, Radio Jordan was rarely heard outside Europe and the Middle East, but the 1995-instituted North American service has been heard around the world with news and pop music.

Kazakhstan

Radio Alma Ata
Zheltoksan St. 175A, 480013, Alma Ata
Prime frequencies: 5035, 5260, 5940, 5960, 5970, 9505 kHz 1830–1900 UTC **EU**

Notes: Radio Alma Ata is rarely heard outside Europe and the Middle East.

Kirbati

Radio Kirbati
P.O. Box 78, Bairiki, Tarawa
Prime frequency: 9825 kHz 0600–1000 UTC

Notes: Radio Kiribati, which broadcasts irregularly from a group of tiny islands in the South Pacific, operates in local languages and English.

Korea (North)

Radio Pyongyang
Pyongyang
Prime frequencies: 9640, 9977 kHz 1500–1550, 1700–1750, 2000–2050 UTC
11335, 13760, 15130 kHz 0000–0050 UTC **NA SA**
15180, 15230 kHz 0400–0450, 0600–0650, 0800–0850 UTC
Also broadcasts in English on: 6576, 9325, 9345, 9640, 11335, 11700, 13650, 13760, 13785, 15340, and 17765 kHz.

Notes: Radio Pyongyang, unlike its fellow communist broadcasters in Laos, Vietnam, and Cambodia, is well heard around the world.

Radio Pyongyang (North Korea)

The following transcription is not from Radio Pyongyang, but it was transmitted via radioteletype on shortwave from North Korea and it is representative of the propaganda that you can still hear via North Korea.

> *Comrade Kim Il Sung, the benevolent father of the people, reading* Believe in the People as in Heaven *as his maxim in the whole period of his revolutionary activities, found himself among the people, shared*

weal and woe with them and made tireless efforts, covering tens of thousands of RI for on-the-spot guidance, crossing high and steep mountain ranges for the freedom and happiness of the people.

The whole life of the respected leader Comrade Kim Il Sung was the most brilliant life of a great revolutionary who forced his way through rigorous storms of revolution, undergoing all sorts of difficulties and trials for the country and the revolution and for the working class and the people and the most glorious life of a great leader who walked a road of victory, turning misfortune into bliss, adversity into prosperity, with his brilliant intelligence, outstanding leadership ability and indomitable will.
— North Korean press agency, KCNA

This propagandist must believe that people will believe anything *so long as you never pause.*

Korea (South)

Radio Korea
18 Yoido-dong, Youngdungpro-gu, Seoul 150-790
02 781 3711
02 781 3799
Prime frequencies: 5975, 6480, 7275 kHz 1900–2000 UTC **AS**
5975, 6480, 9515, 9870 kHz 1600–1700 UTC **AF AS**
6480, 15575 kHz 2100–2200 UTC **EU**
7180 kHz 1200–1300 UTC **AS**
7550, 13670 kHz 0800–0900 UTC **EU**
11810, 15575 kHz 0100–0200 UTC **NA SA**
7205 kHz 0600–0700 UTC **AS**
Also broadcasts in English on: 7275, 9570, 9640, 11740, and 13670 kHz.

Notes: Radio Korea counters the propaganda of Radio Pyongyang with much more laid-back, Western-style programming. The station is heard well around the world because it uses relays in England and Canada, in addition to its own facilities in South Korea.

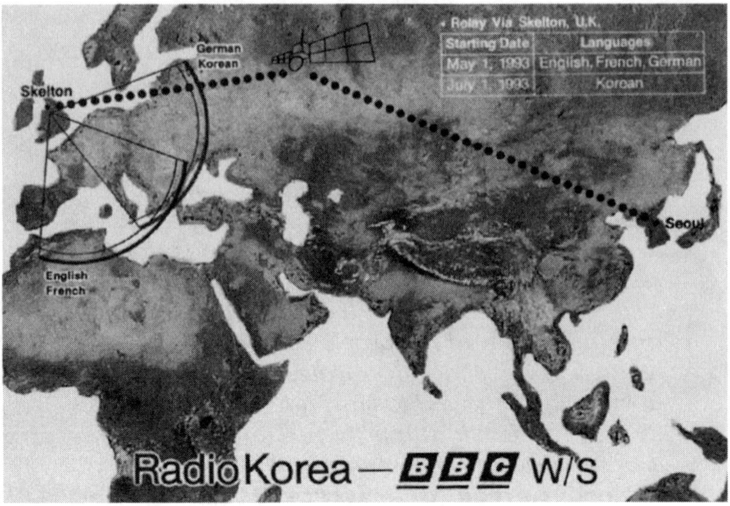

Radio Korea.

Kuwait

Radio Kuwait

P.O. Box 397, 13004 Safat

Prime frequencies: 11990 kHz 1800–2100 UTC **NA EU**

Notes: Radio Kuwait broadcasts some Islamic and political programming, but an hour or more of the the English service each day consists of informal announcers playing a wide variety of Western popular music, including pop, rock, rap, hip-hop, punk, "alternative," and heavy metal.

Kuwait (Radio Kuwait)

- *Kuwait became free on 26 February 1991 with the help of God, the Almighty, and by the will of the Kuwaiti people with the support of the Arab and International Community.*
- *Radio Kuwait embarked on the task of reconstruction after the Iraqi invaders had left the country. All engineers and technicians were working 24 hours/day, non-stop, in order to bring the voice of Kuwait back on air.*

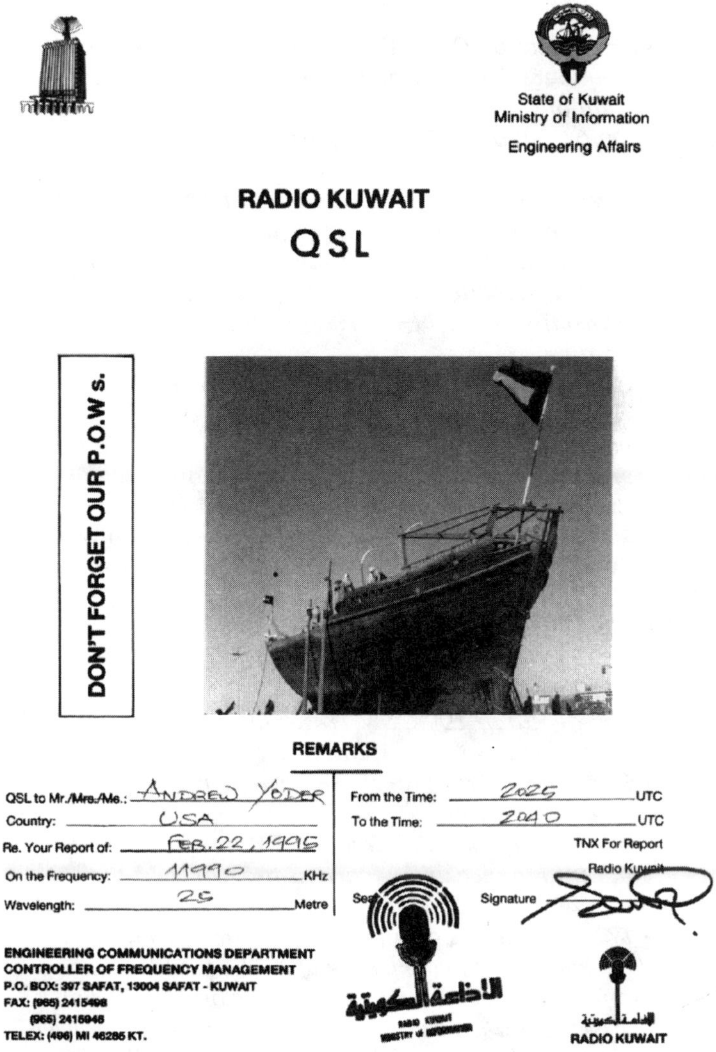

State of Kuwait
Ministry of Information
Engineering Affairs

RADIO KUWAIT

QSL

DON'T FORGET OUR P.O.W s.

REMARKS

QSL to Mr./Mrs./Ms.: Andrew Yoder
Country: USA
Re. Your Report of: Feb. 22, 1995
On the Frequency: 11990 KHz
Wavelength: 25 Metre

From the Time: 2025 UTC
To the Time: 2040 UTC
TNX For Report
Radio Kuwait
Seal
Signature

ENGINEERING COMMUNICATIONS DEPARTMENT
CONTROLLER OF FREQUENCY MANAGEMENT
P.O. BOX: 397 SAFAT, 13004 SAFAT - KUWAIT
FAX: (965) 2415498
(965) 2415946
TELEX: (496) MI 46285 KT.

الاذاعة الكويتية

RADIO KUWAIT

A QSL from Radio Kuwait.

- *The actual radio transmissions began on 3 March 1991 giving Main Arabic program on FM frequencies 97.5 MHz and 87.9 MHz. Afterwards other transmissions went on-air as the reconstruction process gained momentum.*

- *On 5 March 1991, MW transmission on the frequency 540 KHz giving Main Arabic program went on-air.*
- *On 6 March 1991, MW transmission on the frequency 1341 KHz giving Main Arabic program was on-air. Later on various programs were added for transmission on this frequency.*
- *On 1 June 1991, FM transmission on the frequency 98.9 MHz giving Main Arabic program and MW transmission on 1134 KHz giving Main Arabic were on-air.*
- *On 28 August 1991, FM transmission on 92.5 MHz giving English Stereo Music was started.*
- *It is evident that the Radio Kuwait's staff had put in great efforts in bringing its transmissions back on-air within a short period of time.*
- *Subsequently, Radio Kuwait's staff went on to erect SW transmitters and antennae in order to begin the SW transmissions to various targets.*
- *On 17 February 1992, SW transmitter No. 1 went on-air giving Main Arabic program to Gulf, Middle East, Europe and North America.*
- *On 28 March 1992, SW Transmitter No. 2 started its transmissions to cover East and South Asia, Middle East, Europe, and North America giving Main Arabic program.*
- *On 17 October 1992, another FM transmitter was put into service giving various programs on the frequency 100.2 MHz.*
- *Radio Kuwait is committed to give the best quality transmissions to the people of the State of Kuwait and the world as it was the case before the 2 August 1990 and it is striving to provide even better services incorporating the modern broadcasting technology and equipment.*

—Radio Kuwait

Latvia

Radio Latvia International
P.O. Box 266, Riga
Prime frequencies: 5935 kHz, 2030–2100 UTC **EU**

Notes: Radio Latvia International primarily broadcasts a national service in Latvian. The half hour in English seems to have been added as an afterthought.

Lebanon

King of Hope
P.O. Box 3379, Limassol, CYPRUS
Prime frequency: 6280 kHz 0500–2200 UTC

Notes: Operated by High Adventure Broadcasting Ministries (a Christian organization), which also operates the Wings of Hope, Lebanon and KVOH, USA.

Voice of Lebanon
P.O. Box 165271, Al Ashrafiah, Beirut
Prime frequency: 6550 kHz 1800–1815, 2325–2336 UTC

Notes: This government station is sometimes off the air as a result of the political turmoil within Lebanon.

Wings of Hope
P.O. Box 3379, Limassol, CYPRUS
Prime frequency: 11530 kHz 0500–2200 UTC

Notes: Operated by High Adventure Broadcasting Ministries (a Christian organization), which also operates the King of Hope, Lebanon and KVOH, USA.

Lesotho

Radio Lesotho
P.O. Box 552, Maseru 100
Prime frequency: 4800 kHz 0300–2100 UTC

Notes: Broadcasts plenty of local music with announcements in English and other regional languages.

Liberia

ELWA
P.O. Box 10-0192, 1000 Monrovia 10
Prime frequencies: 4760 kHz 0600–2200 UTC

Notes: ELWA is a Christian broadcaster that has operated from Liberia for years. In the midst of political turmoil within the country, its transmitter has been knocked off the air many times. Although the transmitter power has been relatively low for years, the station announced in 1995 that it received a vintage 50-kW transmitter as a gift from FEBC (Philippines).

Liberia Broadcasting System (ELBC)
P.O. Box 594, 1000 Monrovia 10
Prime frequency: 7275 kHz 0550–1000, 1355-1730 UTC
Notes: ELBC is the Liberian government station. During the 1990s revolutions, the transmitter was knocked off the air numerous times, and at one point, two different ELBCs were being operated by rival political factions.

Lithuania

Radio Vilnius
Konarskio 49, LT-2674 Vilnius
370 2 66 05 26
Prime frequency: 9530 kHz 2300–2330 UTC **NA**
Notes: Radio Vilnius, once one of the major outlets of the Soviet Union, has dramatically reduced its output in English.

Malaysia

Voice of Malaysia
P.O. Box 11272, 50740 Kuala Lampur
Prime frequencies: 6175, 9750, 15295 kHz 0555–0825 UTC **AS AUS**
Notes: The Voice of Malaysia is rarely heard outside the South Pacific region.

Malta

Voice of the Mediterranean
P.O. Box 143, Valetta
Prime frequencies: 9765 kHz 0600–0800 UTC
11925 kHz 1400–1600 UTC
Notes: The Voice of the Mediterranean is a joint venture between the Maltese and the Libyan governments, and the programs are broadcast via the Deutsche Welle relay transmitters in Malta.

Moldova

Radio Dniester International
25th October St., 45 Tiraspol Transdniestria, CIS 278000
Prime frequencies: 9620, 11750, 15290 kHz 2030–2100 UTC

Notes: Technically a clandestine station, Radio Dniester International has nonetheless been heard by a number of people outside the European continent. In fact, the signals are apparently better than the official station, Radio Moldova International.

Radio Moldova International
P.O. Box 9972, Chi in u 70, Moldova 277070
Prime frequencies: 7190 kHz 0200–0230 UTC
7235 kHz 1830–1900 UTC
9620 kHz 2130–2200 UTC
15315 kHz 1430–1900 UTC

Notes: The schedules for Radio Moldova International are erratic—the station is often only heard a few times per month.

Monaco

Trans World Radio
BP 349, M-98007, Monte Carlo
33 92 16 5601
Prime frequencies: 7115 kHz 0640–0820, 1130–1200 UTC

Notes: This is one of the many Trans World Radio (a Christian evangelical broadcaster) outlets around the world.

Mongolia

Radio Ulaan Baator
P.O. Box 365, Ulaan Baator
976 1 323096
Prime frequencies: 7290, 12000 kHz 1445–1515 UTC **AS**
7295, 12000 kHz 0330–0400 UTC **AS**
13670, 17900 kHz 1930–2000 UTC **EU**
Also broadcasts in English on: 12015 kHz.

Notes: Mongolia, the land of the Khans, is one of the most mysterious and mythical countries in the world. The country's reputation holds true for its international radio outlet, Radio Ulaan Baator.

Myanmar

Radio Myanmar
GPOB 1432, Yangon
95 1 30211
Prime frequencies: 4725, 5990*, 7185, 9730* kHz 0200–0230, 0700–0730, 1430–1600 UTC

Notes: Rarely heard outside the Indian subcontinent—especially in English.

Namibia

Namibian Broadcasting Corp.
P.O. Box 321, Windhoek 9000
215811
62346
marietha@nbc_hq.nbc.com.na (Internet)
Prime frequencies: 3270 kHz 1800–2200 UTC
3290 kHz 0000–0800 UTC
4930 kHz 1000–1800 UTC
4965 kHz 0800–1800 UTC
Notes: Mostly features light pop music in English.

Nepal

Radio Nepal
P.O. Box 634, Singha Durbar, Kathmandu
Prime frequencies: 3230, 5005 kHz 0000–1700 UTC

Notes: Broadcasts in various languages, including English, with the news in English at 0720 UTC and the "Destination Nepal" travel program at 0850 UTC.

Netherlands

Radio Nederland
P.O. Box 222, 1200 JG Hilversum
31 035 72 42 11
31 035 72 42 52
letters@rn-hilversum.nl (Internet)
letters@rnw.nl (Internet)
Prime frequencies: 5905, 7305, 9860, 11655 kHz 0130–0225 UTC **AS**
6020, 7120, 11655 kHz 1730–1925 UTC **AF**
6020, 6165, 9840 kHz 2330–0125 UTC **NA**

6045, 9560 kHz 1130–1325 UTC **EU**
6165, 9590 kHz 0430–0525 UTC **NA**
9720 kHz 0730–1030 UTC **AUS**
9860, 11655 kHz 0230–0325 UTC **AS**
9895, 13700, 15150 kHz 1330–1625 UTC **AS**
9895, 13700, 15315, 17605 kHz 1930–2025 UTC **AF**
12065, 13705 kHz 1030–1125 UTC **AS**

Also broadcasts in English on: 7260, 9810, 17580, and 21505 kHz.

Notes: Although not as large or well-staffed as some of the other international broadcasters, Radio Nederland has won many awards and is one of the favorites of listeners around the world. In addition to its many interesting features, the station is also home to "Media Network," one of the most popular radio-listening (DX) programs on shortwave.

Radio Nederland.

Netherlands Antilles (Trans World Radio)

The Trans World Radio site at the Netherlands Antilles was once one of the group's largest shortwave outlets. Today, shortwave facilties have been upgraded at the other nine Trans World Radio sites around the world, and the site in the Netherlands Antilles is used for AM and satellite broadcasting.

The stated purpose of Trans World Radio is to assist the Church to carry out the command of Jesus Christ to make disciples of all nations. We do that by using and making available mass communications to (1) preach the gospel to as many new people as possible, (2) instruct those who are already believers in Bible doctrine and daily Christ-like living, and (3) model our message through our own corporate and interpersonal relationships. All programming and operational decisions are made with this purpose in mind.

After much research, it was decided that TWR-Bonaire [Netherlands Antilles] could better use the shortwave money in satellite broadcasts to the cities of Latin America. We became involved in a cooperative venture with HCJB in Quito, Ecuador, to establish a satellite network code named "ALAS (America LAtina via Satellite, which is an acronym for the Spanish word "wings"). We now have at least 23 subscribing stations who receive the signal, choose from a menu of programs being sent via satellite, and broadcast them over their own frequencies. We are working to establish a follow-up system in place in each of those listening areas where people can call for counseling or to get answers to any questions that arise as a result of the programs.

— Bill Early, Program Director, TWR-Bonaire

New Zealand

Kiwi Radio

P.O. Box 3103, Onekawa, Napier

0064 684 48166

Prime frequency: 7445 kHz 0600–0900 kHz (irregular weekends)

Notes: Kiwi Radio has been closed several times by the New Zealand government for broadcasting without a license.

New Zealand (Kiwi Radio)

Under the laws of New Zealand and its people, the entire radio spectrum is a natural resource belonging to the people of New Zealand. It is the responsibility of the Post Office/Broadcasting Tribunal to administer this natural resource for the maximum benefit of the people of New Zealand. We do not believe this has been carried out.

The free radio operators believe that access to the radio spectrum for broadcasting has been effectively closed off to all but those corporations and companies who can provide the enormous capital required to establish a commercial broadcasting station. Such commercial stations, we argue, result in programming aimed for the lowest common denominator of public taste. Such commercial operations shut out many segments of the community from being represented in the programming of commercial stations.

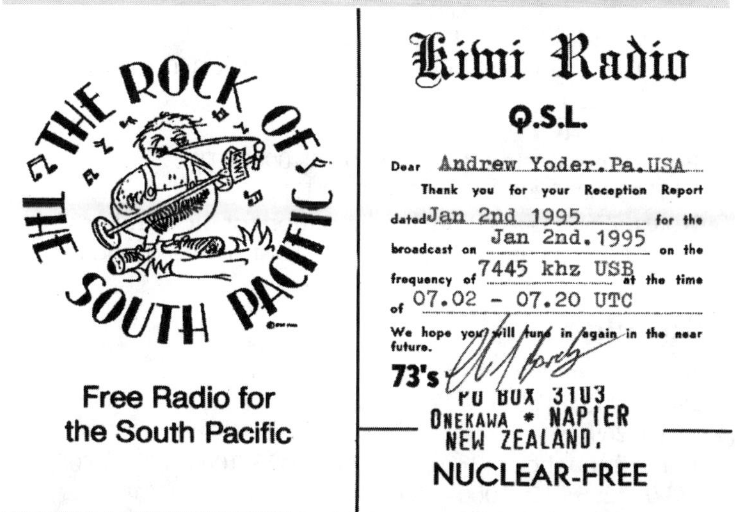

Kiwi Radio/New Zealand

All arguments for the existence of commercial radio are beside the point. The issue is not whether com-

> *mercial stations are truly serving the public, the real issue is why broadcasting should be restricted to commercial interests or publicly supported stations? Why can't an ordinary citizen have a slice of the radio spectrum in which to broadcast?*
> **— Graham Barclay, Kiwi Radio**

Radio New Zealand International
P.O. Box 2092, Wellington
64 4 474 1433
Prime frequencies: 6100 kHz 0717–1206 UTC
9570 kHz 0459–0758 UTC
11910 kHz 1850–2058 UTC
15115 kHz 2051–0458 UTC

Notes: Radio New Zealand International is a small, but popular, broadcaster that lives in the shadow of Radio Australia. In spite of the threats of cutbacks, diehard international RNZI listeners always support the station and help keep it on the air (with letter-writing campaigns, etc.).

ZXLA: Radio for the Print Disabled
First Floor, Levin Shopping Mall, P.O. Box 360, Levin 5500
06 368 2229
06 368 0151
Prime frequencies: 3935 kHz 0600–0900 UTC
3935, 5960, 7290 kHz 1930–0600 UTC

Notes: ZXLA is a national radio station for the blind or vision-impaired. Much of the broadcasts consist of audio books.

Nigeria

Voice of Nigeria
PMB 40003, Falomo, Ikoyi, Lagos
234 1 269 1944
Prime frequencies: 3326, 4990 kHz 0425–1000, 1700–2305 UTC
4990, 7285 kHz 1000–1700 UTC

Notes: The Voice of Nigeria is well heard in many regions of the world and most of its broadcast hours are in English.

Northern Mariana Islands

KFBS
P.O. Box 209, Saipan, CM 96950
Prime frequency: 9465 kHz 1930–2000 UTC
Notes: KFBS is a Christian missionary station in the Far East Broadcasting Company network. Nearly all programming is in a variety of Asian languages. See the feature on Seychelles.

KHBI
P.O. Box 1387, Saipan, CM 96950
Prime frequencies: 9355 kHz 1400–1800 **AF AS**
9355, 9430 kHz 1100–1200 UTC **AS AUS**
9355, 13625 kHz 1300–1400 UTC **AS AUS**
9355, 13770 kHz 1800–2000 UTC **AS AUS EU**
9430, 13615 kHz 0900–1000 UTC **AS AUS**
9430, 13625 kHz 1000–1100 UTC **AS**
13625, 15405 kHz 2200–2400 UTC **AS AUS**
Notes: KHBI is the Pacific branch of Monitor Radio International, the radio voice of the *Christian Science Monitor.*

Norway

Radio Norway International
N-0340 Oslo
47 22 45 8008 (schedule information)
47 22 45 8009 (schedule information)
47 22 45 7134
Prime frequencies: 7480 kHz 0400–0430 UTC **NA**
9590, 11850 kHz 1300–1330 UTC **EU NA**
11850 kHz 1600–1630 UTC **NA**
13800, 15170 kHz 1200–1230 UTC **AS AUS**
15175 kHz 0600–0630, 1600–1630 UTC **AUS**
15220 kHz 0800–0830 UTC **AS**
Notes: These broadcasts are all aired only on Sundays.

Pakistan

Radio Pakistan
Islamabad 44000
92 51 811861
Prime frequencies: 7290, 15190, 17705, 17725, 21730 kHz 0230–0245 UTC **AS**
7485, 11570 kHz 1700–1750 UTC **EU**

9435, 9470, 11570, 13590, 15555, 15675, 17760 kHz 1600–1630 UTC **AF AS**

15625, 17900 kHz 0800–0855, 1100–1120 UTC **EU**

Notes: Features news, sports, commentaries, and readings from the Koran.

Palau

KHBN
P.O. Box 66, Koror, Palau 96940
Prime frequencies: 9830 kHz 1430–1500 UTC
9965 kHz 1230–1300, 1800–1900 UTC

Notes: One of the branches of High Adventure Broadcasting, which also operates the Voice of Hope stations from Lebanon and the United States.

Papua New Guinea

National Broadcasting Commission of Papua New Guinea (NBC)
P.O. Box 1539, Boroko
Prime frequency: 4890 kHz 1900–1400 UTC

Notes: Broadcasts in English and various local languages. This station is the flagship for the network of shortwave broadcasters in Papua New Guinea.

Radio Central
P.O. Box 1539, Boroko
Prime frequency: 3290 kHz 1900–1400 UTC
Notes: Broadcasts in English and various local languages.

Radio Enga
P.O. Box 196, Wabag
Prime frequency: 2410 kHz 1900–1400 UTC
Notes: Broadcasts in English and various local languages.

Radio Milne Bay
P.O. Box 111, Alotau
Prime frequency: 3365 kHz 1900–1400 UTC
Notes: Broadcasts in English and various local languages.

Radio Northern
P.O. Box 137, Popondetta
Prime frequency: 3345 kHz 1900–1400 UTC
Notes: Broadcasts in English and various local languages.

NATIONAL BROADCASTING COMMISSION OF PAPUA NEW GUINEA

Phone: 94 2293 Cables:NATBROADCAST Telex: NE 94 902
FAX 941489

Ref: 16-5-3
Radio New Ireland
P.O. Box 140
Kavieng
New Ireland Province

20th October 1992

Mr Andrew Yonder
P O Box 109
Blue Ridge Summit
PA 17214
United States of America

Dear Mr A Yonder

We write to thank you and acknowledge receipt of your reception report of Radio New Ireland on 27th September, 1992.

Details of the programmes and specification of the frequency mentioned in the report confirmed that you were actually listening to Radio New Ireland.

Radio New Ireland is one of the Provincial Stations of the National Broadcasting Commission of Papua New Guinea. It broadcasts on a frequency of 3905 kHz in the 75 MB and an aerial power of 10KW.

Our transmission times are: 0600 - 0800 GMT
1700 - 2300 GMT

We are pleased to know that Radio New Ireland can be also heard in that part of the world.

Yours faithfully,

Ruben Bale
Provincial Programme Manager

Radio New Ireland from Papua New Guinea.

Radio Western
P.O. Box 23, Daru
Prime frequency: 3305 kHz 1900–1400 UTC
Notes: Broadcasts in English and various local languages.

Philippines

Far East Broadcasting Co. (FEBC)
P.O. Box 1, Valenzuela, Metro Manila 0560
Prime frequencies: 11690 kHz 0930–1100 UTC
11995 kHz 1300–1600 UTC
15450 kHz 0000–0200 UTC

Notes: The FEBC is a Christian missionary station that broadcasts primarily in the more common languages of Asia. The FEBC also has a sister station in the Seychelles.

Radio Pilipinas
Quezon City, Metro Manila 1103
Prime frequencies: 11815, 15190 kHz 1730–1800 UTC
17760, 17865, 21580 kHz 0230–0330 UTC

Notes: Radio Pilipinas, the international broadcast station of the Philippines, is by far the smallest of the four shortwave outlets (Radio Pilipinas, Radio Veritas Asia, FEBC Philippines, and the Voice of America relay station in Poro) in the country.

Pirate broadcasters

"Pirate" stations are unlicensed broadcasters that are primarily hobby operations, although other social and political reasons exist. Pirate stations offer programming that is as varied as the personalities and interests of anyone alive. Some stations feature music (country, rock, punk, hip-hop, rap, big band, reggae, industrial, folk, jazz, blues, international, "alternative," etc.), politics (Liberatian, Democratic, Republican, anarchist, populist, socialist, nazi, etc.) comedy, social views (housing, marijuana laws, welfare, etc.), radio news, telephone calls, etc. Still, pop music and comedic programming are most commonly heard.

Shortwave pirates are very active in North America and Europe, but several "traditional" pirates are also located in other regions (such as Kiwi Radio from New Zealand, Radio G'Day from Australia, and Radio Pirana International from South America).

The traditional frequencies for broadcasting in Europe are 6200–6400 kHz and 7350–7490 kHz. The best frequencies in North America and Australia are 6940–6970 kHz and 7350–7490 kHz. Pirates also occasionally use 1610–1625 (especially in Holland), 9900–9995 kHz, 13950, 15040–15060, and 21455–21500 kHz. For more information on pirate radio, see the books *Pirate Radio: The Incredible Story of America's Under-*

ground, Illegal Broadcasters by Andrew Yoder (HighText, 1996) and *The Pirate Radio Directory* by George Zeller and Andrew Yoder (Tiare Publications, annual).

See also: Ireland and New Zealand

Poland

Polish Radio Warsaw
P.O. Box 46, 00-977 Warsaw
48 22 645 93 05
48 22 444 123
48 22 645 59 17
radio1@ikp.atm.com.pl (Internet)
Prime frequencies: 6000, 6135, 7285 kHz 2030–2125 UTC
6000, 7270 kHz 1800–1855 UTC
9525, 11815 kHz, 1300–1355 UTC
Also broadcasts in English on 7145 and 7270 kHz.

Notes: Formerly known as Radio Polonia, Polish Radio Warsaw is well-heard across much of the world for a few broadcasts in English each day.

Polish Radio Warsaw.

Portugal

Radio Portugal
30-1106, Lisbon
Prime frequencies: 6175, 9570 kHz 0130–0200 UTC **NA**
21515 kHz 1430–1500 UTC **AS**
6130, 9780, 9815, 15515 kHz 2000–2030 UTC **AF**

Notes: Radio Portugual broadcasts with a few half-hour programs in English, but the real focus of the station is its Portugese-language service.

Romania

Radio Romania International
Str. General Berthelot 60-62, P.O. Box 111, Bucharest
401 617 2856
401 312 9262
Prime frequencies: 5990 kHz 0200–0300 and 0400-0430 UTC
6155 kHz 0200–0300 and 0400–0430 UTC
7195, 9570, 11940 kHz 2100–2200 UTC
9510 kHz 0200–0300 and 0400–0430 UTC
9570 kHz 0200–0300 and 0400–0430 UTC
9690, 11810 kHz 1900–2000 UTC
11775, 15250, 17805 kHz 0645–0745 UTC
11830, 15365, 17805 kHz 1730–1800 UTC
11940 kHz 0200–0300 and 0400–0430 UTC

Also broadcasts in English on: 9550, 9665, 9750, 11830, 11775, 11790, 11250, 15335, 15380, 15390, 17250, 17720, and 17790 kHz.

Notes: Despite the fall of communism in Romania, Radio Romania International (formerly known as Radio Bucharest) continues its large, well-heard English service.

Romania (Radio Romania International)

- *Between 1928 and 1944 August 23rd (the beginning of the communist regime in Romania). There was a continuous increase of transmitting stations and broadcasting issues, trying to emancipate the level of development of the country by means of broadcasting issues specially designed to this purpose. Collaborators were even employed—personalities from all domains of activity.*

The English department at Radio Romania International.

- *23rd of August 1944. Bomabardment of the headquarters in Bucharest and Bos Station by the German troops.*
- *1948. Nationalization. The Romanian Broadcasting Society is added to the patrimony of the government.*
- *Building of the Radio House begins, it is partially inaugurated in 1952 and is finished in 1960 when the Concert Hall is also built. Continuous increase of long, medium, short, and ultra-shortwave stations occurs between 1944 and 1948.*
- *Radiodistribution is no longer in charge of the Post and Telecommunication Ministry, but in that of the Radiobroadcasting System.*
- *25th of May 1992. The Audiovisual Law states that the Romanian Broadcasting Corporation becomes a public autonomous service.*
- *14th of July 1993. Romanian Broadcasting Corporation gets the broadcasting license from CNA.*
- *Romanian Parliament adopts the Law on the operation of the Romanian Radio Corporation and the Romanian Television Corporation.*

— Radio Romania International

Russia

Voice of Russia
ul. Pyatnitskaya 25, 113326 Moscow

Prime frequencies: 5920 kHz, 5995, 6055, 7180, 7400, 9550, 9890 kHz 2000–2300 UTC **AF EU**
5930, 7345, 9895 kHz 0500–0800 UTC **NA**
6000, 6060, 7160, 7205 kHz 1200–1500 UTC **AS**
7105, 7165 kHz 0000–0200 UTC **NA**
7105, 7165, 7180, 7270, 9850, 12050 kHz 0200–0600 UTC **NA**
7115, 9600, 12025, 15265 kHz 1500–1700 UTC **AS**
7165, 12015, 12065, 15230, 15480, 21530 kHz 1300–1600 UTC **AS**
7170, 7180, 7205, 9550, 9860, 9890 kHz 1600–2100 UTC **AF EU**
9550, 9800, 11675, 11710, 12015 kHz 0900–1300 UTC **AUS**
9550, 15140 kHz 1300–1500 UTC **AUS**
15230, 17620, 17840 kHz 0600–0900 UTC **AUS**

Also broadcasts in English on: 4740, 4795, 5940, 5965, 5975, 6015, 6085, 6185, 7130, 7135, 7150, 7180, 7185, 7195, 7205, 7250, 7295, 7300, 7325, 7335, 7380, 7420, 9510, 9550, 9620, 9750, 9835, 9875, 9885, 11655, 11675, 11760, 12015, 12035, 12040, 15140, 15150, 15160, 15295, 15425, 15455, 15460, 15470, 15560, 15580, 17570, 17580, 17675, 17720, 17755, 17890, 21600, 21790 kHz.

Notes: Once one of the very largest shortwave broadcasters in the world, the Voice of Russia (formerly known as Radio Moscow) has cut its number of personnel nearly in half. But with more than 1200 employees, it is still enormous, and the station can be heard around the world.

Russia (Voice of Russia)

Our main requirement to broadcasting is not to give one-sided coverage or advertise just one point of view. We insist that the entire range of opinions, views, and platforms be represented. In this sense, we are something like a mirror, reflecting the complicated and controversial processes underway in Russia.

Naturally, in foreign policy, we, above all, present the views of the state. As an official broadcaster, the Voice of Russia defends Russia's interests in the same way that the Voice of America, the BBC, and other

Russia

national broadcasting stations speak for their countries. People want to know the Kremlin's stand on various problems.

Our main task is to tell people about Russia, make it closer and easier to understand. Not to create a false image of an Emerald City, but to project a true and recognizable image. We have taken a big step away from ideology and have revised our views of the world much more radically than the world has revised its views of us.

— Armen Oganessian, Chairman, Voice of Russia

Rwanda

Radio Rwanda
Kigali
250 7 6185

Prime frequencies: 6055 kHz 0515–0525, 1145–1155, 1915–1925 UTC

Notes: Programs are via the facilities of the Deutsche Welle in Rwanda.

Serbia

Radio Yugoslavia
Hilanarska 2, 11000 Beograd
Prime frequencies: 6100, 6185 kHz 2100–2130 UTC **EU**
9580 kHz 0000-0030, 0430–0500 UTC **NA**

Notes: The United Nations has built up troops in the regions in and around the former country of Yugoslavia in an attempt to prevent Serbia from invading the rest of Croatia and Bosnia. Because most of the world is in either the United Nations or is Islamic, most media outlets are pro-Croatian or Bosnian. For news and commentaries from a Serbian perspective, one of the only outlets is Radio Yugoslavia, which is widely heard.

Seychelles

FEBA Radio
P.O. Box 234, Mahe
Prime frequency: 15555 kHz 0500–0545 UTC **AS**

Notes: FEBA Radio in Seychelles is a Christian missionary sister station to the FEBC in the Philippines.

Seychelles (FEBA Radio)

FEBA is an evangelical interdenominational mission-motivated broadcasting organization, supported by Christian groups, mainly in Britain, but also in Canada, Singapore, Australia, and New Zealand. It provides facilities for churches and Christian organizations in the listening areas to enhance their ministry by enabling them to speak to the people in their own language by shortwave radio.

FEBA's transmitting base is in the Seychelles, islands situated between the northeast of Madagascar and the equator. The studios and administration are at Sans Souci, a village 250 meters high, overlooked by the main mountain of Mahe, Morne Seychellois. The transmitting station is at Anse Etoile on the northeast coast. The aerial system is built on a submerged coral reef, about a kilometer offshore.

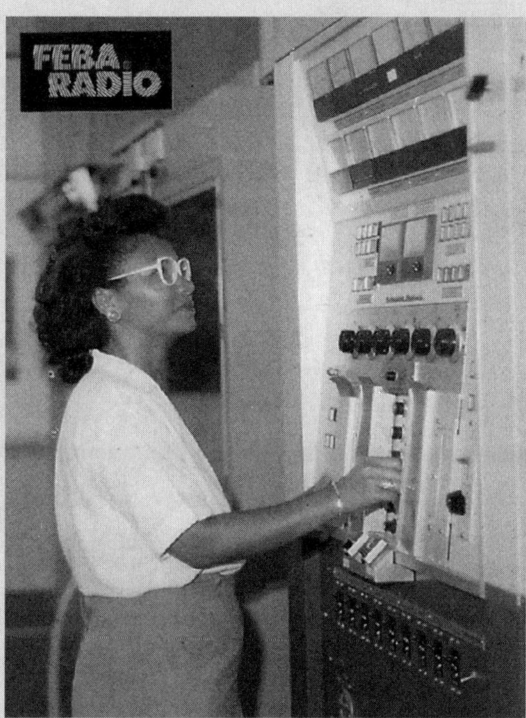

Seychelles

The Seychellois (pronounced "Say-shel-wah") are descendants of French and Asian settlers, and freed African slaves. Most of the present-day Seychellois are a delightful mixture of these three ethnic groups, although some retain one or other distinctive feature. They are warm, friendly, and follow a simple lifestyle.
— FEBA

Sierra Leone

Sierra Leone Broadcasting Service
New England, Freetown
Prime frequencies: 3316, 5980 kHz 0600–2230 UTC

Notes: Sierra Leone is not widely reported outside Northern Africa and southern Europe.

Sierra Leone (SLBS)

The Sierra Leone Broadcasting Service (SLBS) was inaugurated in 1934 as the Freetown Rediffusion Service—the first radio service in Anglophone West Africa, by the late Sir Arnold Hudson (then the British Colonial Governor of Sierra Leone). Since 1934, sound broadcasting in Sierra Leone has evolved from a rediffusion service (Kongosah Box/Gossip Box) to a complete wireless network.

Sierra Leone

The national service of the SLBS maintains an on-air presence for approximately 12 hours weekly and broadcasts mainly in English, Mende, Tenme, Limba, and Krio. Although Sierra Leone does not have a broadcasting act, the SLBS adheres strictly to an unwritten code of ethics, which guarantees freedom of expression governed by a keen sense of social responsibility.

As a government institution, it also maintains a policy of public service broadcasting: to inform, educate, and entertain, as well as mobilize the population for national development. Its program service offers listeners a variety of material, including news and current affairs, features, women and children, health, sports, music, and religious (Christian and Islamic) programs.
— SLBS

Singapore

Radio Singapore International
Farrer Rd., P.O. Box 5300, Singapore 9128
65 3535300
65 2591380
Prime frequencies: 6155, 9530 kHz 1100–1400, 2300–2400 UTC

Notes: RSI features a number of programs, including Dateline RSI, Chartbeat, and "You Asked For It" with plenty of local musicians from Singapore.

Slovakia

Radio Slovakia International
Mytna 1, 812 90 Bratislava
042 7498 075
042 7496 282
Prime frequencies: 5915, 7345 kHz 1930–2000 UTC **EU**
5930, 9440 kHz 0100–0130 UTC **NA**
11990, 17485, 21705 kHz 0830–0900 UTC **AUS**

Notes: Listening to Radio Slovakia International is like a mini-tour of Slovakia. Instead of focusing on the world (like the BBC) or on a region (like Radio Australia), almost everything on RSI pertains to Slovakia. Some of the programs include Slovak Kitchen, Back Page News, Slovak Personalities, and Listeners' Tribune.

Slovak Radio.

Solomon Islands

Solomon Islands Broadcasting Corp. (SIBC)
P.O. Box 654, Honiara
677 23159
Prime frequencies: 5020, 9545 kHz 0500–1100, 1900–2200 UTC

Notes: The SIBC is a tiny local broadcast service for the Solomon Islands, broadcasting in English and Pidgin. The station typically features island music, news, and public service announcements, but it is not heard well outside the South Pacific.

South Africa

Channel Africa
P.O. Box 91313, Auckland Park 2006
Prime frequencies: 3220, 5955 kHz 0300–0500 UTC
3220, 7240 kHz 1500–1800 UTC
5955, 9695 kHz 0500–0600 UTC
9695 kHz 1600–1700 UTC

News: Channel Africa used to be known as Radio South Africa and Radio RSA and was the largest shortwave broadcaster in Africa. Today, the name is different and the hours on the air have been slashed.

Sentech Shortwave Services
Private Bag X06, Honeydew 2040
475-5112
Prime frequencies: 3320, 4810 kHz 2300–0300 UTC
7270 kHz 0500–1430 UTC
9630 kHz 0800–1630 UTC

Notes: These channels all feature programs from different branches of the South African Broadcasting Corporation (Radio 2000, Radio Oranje, Afrikaans Stereo, etc.). However, the transmitters are all operated by the Sentech company.

Spain

Radio Exterior de Espana
(Spanish Foreign Radio)
Apartado 156.202, 28080 Madrid
346 11 49
346 18 15
Prime frequencies: 6125 kHz 2100–2200 UTC **EU**
9540 kHz 0000–0200, 0500–0600 UTC **NA**
9675 kHz 1900–2000 UTC **AF**

Notes: Like its peninsula "roommate" Portugal, Radio Exterior de Espana has a limited English-language output and instead concentrates its efforts on its home language.

Spain (REE)

The first news of Radio Nacional de Espana broadcasts to the outside world came in the magazine Radio Nacional *on March 8, 1942. It announced that the Deputy Secretary for the Popular Education had organized new radio broadcasts that Radio Nacional was to begin on March 15, 1942.*

Throughout its existence, what we now call Radio Exterior *has gone through many "transformations and changes, just as the rest of Spanish National Radio has, as an immediate consequence of Spain's recent history." Mr. Candau states that if in its origins, Radio Exterior was the propaganda platform for a dictatorship, today it is the voice of "a new Spain, young and dynamic, free and democratic."*

RADIO EXTERIOR DE ESPAÑA

50 AÑOS EN LAS ONDAS DEL MUNDO

Spain

Diego Carcedo, Director of Radio Nacional de Espana, also remarked on the origins of REE in his reflections on 50 years of broadcasts. He stated that when the broadcasts were created in the midst of World War II, they were begun as a platform to the world for a "dictatorial regime which suffocated the freedom and voices inside of Spain," adding that everything points to the foreign broadcasts having been created for the "undeclared purpose of presenting to the international community an image of Francoism which, without denying its 'peculiarities' could inspire benevolence." Now, REE's ideological aims are "radically opposed" to those that inspired the international broadcasts of Franco's time. Trying to "legitimize the illegitimate" has given way to a "decisive defense of democratic values, of human rights, and, in short, of human aspiration to freedom."

— Pilar Salvador, Relaciones con la Audiencia, REE

Sri Lanka

Sri Lanka Broadcasting Corporation
P.O. Box 574, Colombo 7
Prime frequencies: 7190, 11800 kHz 0030–0430, 0630–1030, 1330–1730 UTC

9720 kHz, 15425 kHz 0030–0430, 1230–1730 UTC **AS**
9720, 15425 kHz 0445–0515 UTC **NA**
Also broadcasts in English on: 6005, 6075, 15120, and 17850 kHz.
Notes: The SLBC is not widely heard outside Asia.

Swaziland

Trans World Radio
P.O. Box 64, Manzini
268 55333
Prime frequencies: 3200 kHz 1745–2045 UTC
3240 kHz 1900–2045 UTC
5055, 6070, 9500 kHz 0430–0835 UTC
9500 kHz 1600–1830 UTC
Notes: One of the many Trans World Radio (a Christian evangelical broadcaster) outlets around the world.

Sweden

Radio Sweden
S-105 10 Stockholm
46 8 784 50 00
46 8 667 62 83
46 8 667 37 01 (schedule via fax)
ftp://town.hall.org/radio/Mirrors/RadioSweden/ (to download Media Scan program from Internet)
Prime frequencies: 6065, 6200 kHz 0030–0100 **SA**
6065, 9655 kHz 1830-1900, 2130–2200 UTC **AF AS**
6065, 9655 kHz 2130-2200, 2230–2300 UTC **EU**
6200, 9850 kHz 0230-0300, 0330–0400 UTC **NA**
9895, 11695 kHz 0130–0200 UTC **AS AUS**
11650, 15240 kHz 1330–1400, 1430–1500 UTC **NA**
13775, 15120, 15240 kHz 1230–1300 UTC **AS AUS**
Also broadcasts in English on: 11910 and 13690 kHz.
Notes: The ever-popular Radio Sweden broadcasts such programs as Upstream, In Touch With Stockholm, Sounds Nordic, Media Scan, and Spectrum.

Sweden (Radio Sweden)

We are often asked how many people hear our programs. Unfortunately, there's no way to give an honest answer. All we can say is that the 40,000 letters we receive from listeners every year compares well with stations of the same size in other countries.

Some of the major international broadcasters, such as the BBC, have tried to calculate the size of their audience based on the number of letters received. However, the BBC has now rejected this method. After carrying out the first representative listener surveys in Hungary and Poland, the BBC discovered that previous audience estimates were far exaggerated.

Despite this experience, listener reports are important to us at Radio Sweden. They give us a reliable indication where in the world we can be heard well, or poorly. This allows us to make adjustments, such as change to more suitable frequencies.

The letters are also an inspiration for program producers and evidence that there is interest abroad in Sweden.
— Radio Sweden

Switzerland

Swiss Radio International
Giacomettistrasse 1, 3000 Berne 15
031 350 92 22
031 350 95 44
Prime frequencies: 5885 kHz 0100–0130 UTC **NA**
6135 kHz 0100–0130, 0400–0430, 2000–2030 UTC **AF NA**
9770, 11640 kHz 2000–2030 UTC **AF**
9885 kHz 0100–0130, 0400–0430, 0600–0630, 1700–1730 UTC **AF NA**
9905 kHz 0100–0130, 0400–0430 UTC **NA**
13635 kHz 1100–1130, 1700–1730, 2000–2030 UTC **AF AS**
13685, 17515 kHz 0900–0930 UTC **AS AUS**

Also broadcasts in English on: 6165, 6205, 7250, 7480, 11640, 11650, 12075, and 15340 kHz.

Notes: Switzerland is known for impartiality in politics, so be sure to check out the world features (NewsNet, World Scene, Swiss Scene, and Down To Earth) on Swiss Radio International.

Radio Switzerland. A calm moment in the news department.

Syria

Radio Damascus
Ommayad Sq., Damascus
Prime frequencies: 12085, 15095 kHz 2010–2105 UTC **EU**
12085, 15095 kHz 2110–2210 UTC **AS AUS NA**

Notes: Radio Damascus is a great source of information on the Middle East through such programs as Welcome to Syria, Arab Civilization, Arab Profile, Syria & the World, and Arab Newsweek.

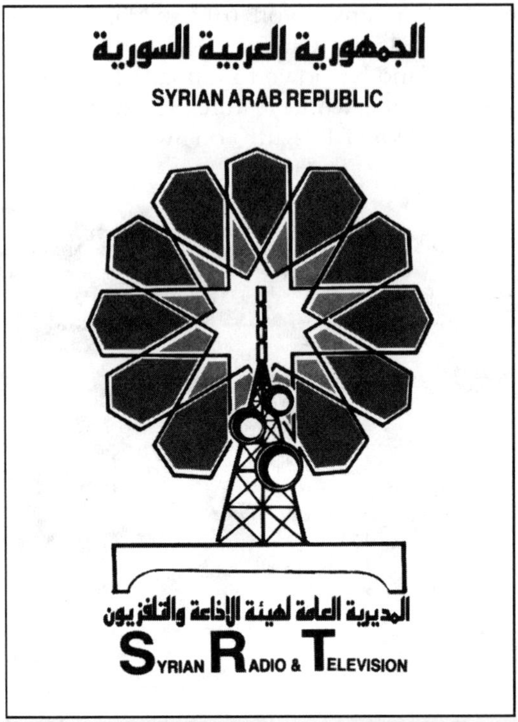

Radio Damascus.

Tajikistan

Tajik Radio
31 Chapyev St., Dushanbe
Prime frequency: 7245 kHz 0345–0400, 1645–1700 UTC **AS**
Notes: Tajik Radio is rarely heard outside the Middle East.

Taiwan

Voice of Free China
53 Jen Ai Rd., Sec. 3, Taipei
☎ 02 771 0150
Fax 02 751 9277
Prime frequencies: 5810 kHz 2200–2300 UTC
5950 kHz 0200–0400, 0700–0800 UTC
7130 kHz 0200–0300, 1200–1300 UTC
9680 kHz 0200–0400 UTC

11825 kHz 0200–0400 UTC
15345 kHz 0200–0400 UTC

Also broadcasts in English on: 9610, 9850, 11740, and 11745 kHz.

Notes: The Voice of Free China is one of the major shortwave outlets from Asia. In addition to providing information about the views and culture of Taiwan, the VOFC even informs its listeners with newspapers and station mini magazines.

自由中國之聲
VOICE OF FREE CHINA

The Voice of Free China
P.O. Box 24-38
Taipei, Taiwan
The Republic of China

TEL: 886-2-771-0151 ext. 2424
FAX: 886-2-771-0151 ext. 2405
886-2-751-9277

Dear Listeners,

It's the time for us to invite you to join the fun for celebrating the Chinese New Year again!

This year we are going to ask you to be part of our family on the air, that is, you can join in the family reunion time by telling us some of the crazy things, funny jokes, or heart-warming stories of your life. And you can do this by either sending in a recorded cassette, or by writing out the story in a letter, and enclosed at least a few words about how you would celebrate the Chinese New Year if you were in Taiwan.

The idea is we would like to get to know you more, our listeners, as we celebrate the family reunion time on the air. Instead of gathering around a fire place as a family, this year, we can each cling to a radio set listening to our family members talking about what has happened in their lives. I am sure this would prove to be a distinctly different experience as you listen to stories shared by your family members on the air coming from all corners of the world.

We feel excited already as we envision what great fun we are going to have. Please, respond speedily. We can't wait to find out whom we've been speaking to on the air all these years!

Reply ASAP!!

Love,

Wendy Ju, Philip Wang

Note: This year's Chinese New Year falls on **January 31, 1995**.
So, send in your story before **January 10.**
Don't let us wait too long!!!

Voice of Free China from Taiwan.

Thailand

Radio Thailand

236 Vibhavaadi-Rangsit Rd., Din Dang, Kuay-Khwang, Bankok 10400

2776139

Prime frequencies: 9555 kHz 1900–2000 UTC **EU**

9580 kHz 0000–0030 UTC **AS AF**

11835 kHz 2030–2045 UTC **EU**

11890 kHz 0300–0330 UTC **NA**

11905 kHz 0030–0100 UTC **NA**

Notes: Although Radio Thailand is an international broadcaster, its programs cover regional events (such as concerts, weather, etc.) and it often sounds more like a national service.

Turkey

Voice of Turkey

P.O. Box 333, 06.443 Yenesehir, Ankara

490 98 00

490 98 11

490 98 45

490 98 46

Prime frequencies: 9400 kHz 2100–2200 UTC

9445 kHz 2300–2400, 0400–0500 UTC

9675 kHz 1330–1400

11710 kHz 2300–2400

Notes: With such program names as A Haven in the East: Turkey and Magnificent Istanbul, there is no question that the Voice of Turkey is playing up its role as a tourist lure.

Turkey (Voice of Turkey)

Winners of the 1994 Voice of Turkey essay contest could win a 10-day vacation to Turkey. The subject of the essay was "How can friendly relations between our countries be developed further and what is the role of the media in this?" Here are a few segments of the winning responses.

> *As long as cultures which respect all individuals and values continue to exist, hopes for justice, cooperation, tolerance, and righteousness will increase and one day the world will be restructured on the basis of*

Turkey

these values . . . (I think, however, that) the media should voice not only culture and the voice of its own nation and culture but also the voice of the other nations and cultures.

— Professor Dr. Ferid Muhic, Macedonia

Unfortunately no one has the chance to visit all of the countries in the world and get to know their people and their cultures. Instead, in this matter we rely on second-hand information that we receive from newspapers, television, and radio . . . The existence of these instruments of mass media are important: the media must be able to transform the xenophobia and the anxiety felt about strangers and the cultures, into curiosity.

— Karl Heinz Sahlmann, Germany

To be honest, I did not know much about Turkey until I saw a number of documentaries on the television and read articles about Turkey in periodicals and newspapers . . . The role of radios in improving friendly relations between countries is incontestable. For instance, if it was not for the Persian-language broadcasts of the VOT, Iranians could not have participated in this essay contest.

— Nesrin Puyanzade, Iran

Ukraine

Radio Ukraine International
vul Kreshchatik 26, 252001 Kiev
Prime frequencies: 6055, 7180, 9810, 11870 kHz 0100–0200, 0400–0500 UTC
5940, 6020, 7180, 7240, 11870 kHz 2200–2300 UTC
Notes: Radio Ukraine International (formerly known as Radio Kiev) is one of the only radio stations from the former Soviet Union that delivers solid signals to much of the world. The broadcasts are quite Western—what a surprise to hear a speed metal song by Metallica during one RUI program!

United Arab Emirates

UAE Radio from Abu Dhabi
P.O. Box 63, Abu Dhabi
971 2 451155
Main frequencies: 9605 kHz 2200–2400 UTC
9770 kHz 2200–2400 UTC
11710 kHz 2200–2400 UTC
Notes: UAE Radio from Abu Dhabi is operated by the UAE Ministry of Information and Culture.

UAE Radio Dubai
P.O. Box 1695, Dubai
Prime frequencies: 13675, 15320, 15395 kHz, 1030–1110, 1330–1400, 1600–1640 UTC
21605 kHz 1030–1110, 1330–1400 UTC
Notes: UAE Radio from Dubai, like the station from Abu Dhabi, is well heard across most of the world with news, Arabic music and culture, and Koran readings.

United Kingdom

British Broadcasting Corp. (BBC)
44 171 257 2211
44 171 257 8258
44 171 240 4899
iac@bbc-ibar.demon.co.uk (Internet)
BBC Worldwide program schedule available from:
P.O. Box 3000
Denville, NJ 07834
USA

☎ 201 627 2427
📠 201 627 5827

Prime frequencies: 3255 kHz 0300–0600, 1615–2200 UTC **AF**
5970 kHz 0000–0330 UTC **SA**
5975 kHz 2100–0600 UTC **NA**
6005 kHz 0300–0730, 1700–2200 UTC **AF**
6175 kHz 2300–0430 UTC **NA**
6180 kHz 0300–0730, 1700–2230 UTC **EU**
6195 kHz 1700–0200, 0400–0630, 0900–1615 UTC **EU AS NA**
7180 kHz 1300–1745 UTC **AS**
7325 kHz 2300–0230 UTC **NA SA**
9410 kHz 0000–2230 UTC **EU AS**
9515 kHz 1100–1715 UTC **NA**
9600 kHz 0300–0800 UTC **AF**
9605 kHz 0100–0430 UTC **AS**
9630 kHz 1700–2200 UTC **AS**
9640 kHz 0500–0815 UTC **NA**
9740 kHz 0500–1615 UTC **AS**
9915 kHz 2200–0330 UTC **SA**
11750 kHz 0900–1800 UTC **AS SA**
11760 kHz 0300–1400, 1700–2200 UTC **EU AS**
11835 kHz 1930–2315 UTC **AF**
11865 kHz 1200–1615 UTC **NA SA**
11955 kHz 0000–0930, 1100–1300 UTC **AS**
12095 kHz 0400–2230 UTC **EU**
15070 kHz 0600–2130 UTC **EU**
15280 kHz 0330–1030 UTC **AS**
15310 kHz 0300–1500 UTC **AS**
15360 kHz 0000–0330, 0500–0800, 1100–1300 UTC **AS**
15400 kHz 0600–1130, 1430–1930, 2030–2130 UTC **AF EU**
15575 kHz 0400–1500 UTC **EU AS**
17640 kHz 0400–1500 UTC **EU AS**
17705 kHz 0900–1615 UTC **EU**
21470 kHz 1400–1630 UTC **AF**

Also broadcasts in English on: 3915, 5965, 6135, 7110, 7160, 7250, 9510, 9575, 9580, 9590, 9610, 9740, 9760, 11680, 11765, 11775, 11860, 11945, 15190, 15380, 17790, 17830, 17880, 21490 kHz.

Notes: If it's happening in the world, chances are that you'll hear about it on the BBC.

President Moi of Kenya spoke to Anna Horsbrugh Porter for the BBC World Service series "The Legacy of Colonialism."

United Kingdom (BBC)

No literature from the station and no comments from the station's English department can adequately describe the BBC. This is not to say that the BBC's public relations department is lacking, but simply that they are the best and the biggest, and they don't need to tell people because everyone already knows it.

Ever since World War II, the BBC has been renowned for its unbiased worldwide news coverage. The BBC's news coverage spans the globe—BBC reporters trek across every country for a story and the English-language programs are beamed on many frequencies to every continent. Even though the station covers the world, that coverage isn't "thin." Often, the BBC coverage of international news events are more in-depth than those within the country that the event occurred in. As a result of this spectacular news coverage, many people purchase shortwave radios with the sole purpose of listening to the news on the BBC.

But the BBC isn't just news and news-related programs. Every year, the BBC wins by a landslide in the polls and ratings of international broadcast stations. Of course, part of this is because of the news coverage, but the BBC features so many different programs that they produce a monthly magazine-formatted schedule that covers the different shows that will be aired over the upcoming month. For Americans, this guide is somewhat like a TV Guide *for one shortwave station. The guide features numerous programs that fit under the following headings: Arts, Background current affairs, Business and finance, Current affairs, Classical music, Drama, Development/environment, Education, General features, Human interest, Light entertainment, Media, News, Popular music, Religion, Readings, Science and technology, Sports, and Youth.*

USA

KAIJ
P.O. Box 270879, Dallas TX 75227
817 277 9929
Prime frequency: 5810, 9815 kHz 0200–1400 UTC
13740, 13815 kHz 0000–0200 UTC
Notes: KAIJ features plenty of the end-times preaching of Dr. Gene Scott.

KJES
Star Rt Box 300, Mesquite, NM 88048
714 731 4196
Main frequencies: 11715 kHz 1300–1600 UTC **NA**
15385 kHz 1800–1900 and 2000–2100 UTC **NA AUS**
Notes: KJES is a refreshingly simple Christian outreach broadcaster. It features no promotional material, just a straightforward message.

KNLS
Anchor Point, AK 99556
615 371 8791
Prime frequencies: 7365 kHz 1300–1400 UTC
9615 kHz 0800–0900 UTC
Notes: KNLS (The New Life Station) targets listeners in Russia and the Pacific region. However, it does have several English-

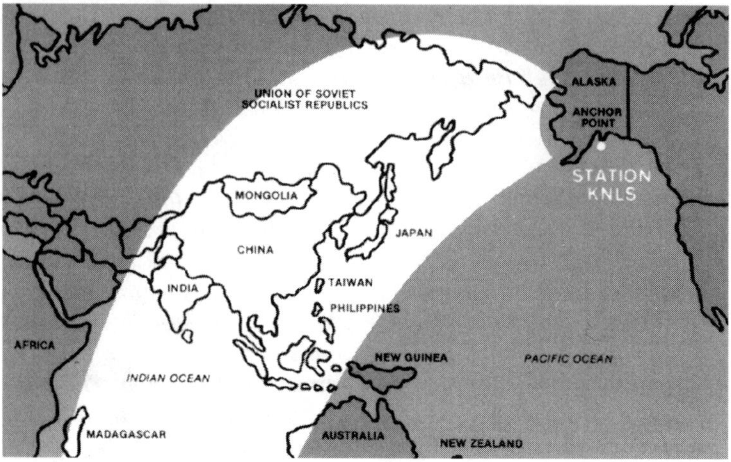

KNLS from the United States.

language broadcasts and features an excellent DX program in addition to its primary focus of Christian outreach.

KTBN
P.O. Box A, Santa Ana, CA 92711
714 730 0661
Prime frequencies: 7510 kHz 0000–1600 UTC **NA**
Notes: Features broadcasts of conservative Christian preachers on the Trinity Broadcasting Network.

KVOH
P.O. Box 93937, Los Angeles, CA 90093
Prime frequencies: 9785 kHz 0200–0700 UTC **NA SA**
17775 kHz 1200–2400 UTC
Notes: Operated by High Adventure Broadcasting Ministries (a Christian organization), which also operates the Wings of Hope, Lebanon and KVOH, USA.

KWHR
Naalehu, HI
Prime frequencies: 6120 kHz 1600–1800 UTC **AUS**
9930 kHz 0800–1600 UTC **AS**
13625 kHz 1800–2000 UTC **AUS**
15405 kHz 2000–2200 UTC **AS**
17510 kHz 0000–0400 UTC **AS**
17780 kHz 0400–0800 UTC **AS**

Notes: KWHR is the Pacific sister station of WHRI (World Harvest Radio).

Voice of America
330 Independence Ave. SW, Washington, DC 20547
202 401 1493
202 401 1494
letters@voa.gov (Internet from outside the US)
letters-usa@voa.gov (From within the US)

Prime frequencies: 3980, 6040, 9760, 15205, 19379L kHz 1700–2000 UTC **AF EU**

3985, 5995, 6873L, 7170, 15205 kHz 0400–0630 UTC **AF EU**
4875, 11920, 13710, 15410, 17895 kHz 1800–1900 UTC **AF**
5980, 7105, 7405, 9575, 15300 kHz 0300–0430 UTC **AF**
5985, 11720, 15425 kHz 1000–1100 UTC **AS AUS**
5990, 6045, 7150, 9645, 9690, 11870 kHz 1700–1800 UTC **AS AUS**
6035, 7405, 9630, 12080 kHz 0500–0630 UTC **AF**
6040, 6160, 9760 kHz 2100–2200 UTC **AS**
6040, 9700, 9760 kHz 1800–2100 UTC **AS**
6040, 9760, 15205, 19379L kHz 2000–2200 UTC **AF EU**
6110, 9645, 9760, 15160, 15425 kHz 1100–1500 UTC **AS AUS**
6110, 7125, 9645, 15255, 15395 kHz 1400–1800 UTC **AS**
6110, 7215, 9645, 9760 kHz 1500–1600 UTC **AS**
6130, 7405, 9455, 11695, 13740, 15205 kHz 0000–0200 UTC **NA SA**
6140, 11965 kHz 0600–0700 UTC **AS**
6165, 7405, 9590 kHz 1000–1200 UTC **NA SA**
7115, 9635, 11705, 15170, 17740, 21550 kHz 0100–0300 UTC **AS**
7215, 9705, 11760, 15185, 17735 kHz 2200–2400 UTC **AS AUS**
7215, 9770, 11760, 15185, 17735 kHz 0000–0100 UTC **AS AUS**
7265, 9575, 15300 kHz 0430–0500 UTC **AF**
7340, 7375, 7415 kHz 2200–2230 UTC **AF**
7375, 7415, 11920, 12040, 15410 kHz 1900–2000 UTC **AF**
7375, 7415, 15410, 15445, 17725 kHz 2000–2230 UTC **AF**
9525, 11870, 15180 kHz 1900–2000 UTC **AS AUS**
9670, 15155 kHz 2200–2300 UTC **NA SA**
9700, 15205 kHz 1500–1800 UTC **AS**
11870, 15185, 17735 kHz 2100–2200 UTC **AS AUS**
11920, 13710, 15410, 15445, 17895 kHz 1600–1730 UTC **AF**
11965, 15205 kHz 0400–0600 UTC **AS**

Notes: The Voice of America is the official voice of the United States Information agency. It represents the views of the US government and airs plenty of popular broadcasting features from locations all over the world.

United States (Voice of America)

The international shortwave broadcasters are in a constant state of flux; some stations are destroyed in wars, some expand their services for political reasons or to show the world that their nation is emerging technologically, some cut back to experiment with other fields of communications (such as satellite radio and TV), some governments are ignorant of the service that their international service provides and the broadcasts are lost admidst budget cutbacks. Here's a portion of the official press release regarding the closure of one of the Voice of America relay stations.

BETHANY RELAY STATION
VOICE OF AMERICA

September 23, 1944
November 14, 1994

Voice of America

Washington, DC, 16th November: The United States Information Agency's Voice of America Bethany relay station ceased international shortwave broadcasting on 14th Novemeber 1994.

The International Broadcasting Act of 1994 provides for cuts of $400,000,000 over the next four years and *for the consolidation of all US overseas broadcasting [Voice of America, Radio Free Europe/Radio Liberty, Radio and TV Marti, and Radio Free Asia]. The consolidation has meant a realignment of US facilities used to broadcast shortwave, including the closing of the Bethany relay station. By 1st January 1995, USIA will report the Bethany property to the General Services Administration (GSA) as excess to its needs.. . . GSA applies priority guidelines to determine what becomes of excess property. The first step is to determine if any other federal agency has a requirement for the property. If no other federal uses are determined, alternative uses include assistance to the homeless, health and educational facilities, parks and recreational facilities, correctional facilities and wildlife conservation.*

With Bethany off the air, the site could soon see more relaxed days as a wildlife reserve or as a prison.

WEWN
P.O. Box 100234, Birmingham, AL 35210
Prime frequencies: 5825 kHz 0000–0200 UTC **EU**
7425 kHz 24 hrs **NA**
9540 kHz 1200–1300 UTC **SA**
11820 kHz 2100–2200 UTC **EU**
13695 kHz 1800–1900 UTC **EU**
15375 kHz 0000–0100, 1800–2100, 2200–2300 UTC **SA**
15695 kHz 1100–1300 UTC **AS**

Notes: Aside from Vatican Radio, WEWN is one of the only English-language Catholic stations in the world. The station features everything from Catholic preachers to call-in Catholic question talk shows.

WEWN from the United States.

WHRI
P.O. Box 12, South Bend, IN 46624
219 291 8200
219 291 9043
Prime frequencies: 7315, 7355 kHz 0800–1000 UTC
7315, 9580 kHz 1000–1300 UTC
7315, 9495 kHz 2300–0800 UTC
9465, 15105 kHz 1300–1700 UTC
9485, 13760 kHz 1800–2100 UTC
13760, 15105 kHz 1700–1800 UTC
13760, 17830 kHz 2100–2300 UTC

Notes: Broadcasts various programs in English and Spanish. In addition to the somewhat Christian content of WHRI, the station also relays plenty of programs from anti-Castro clandestines for audiences in Cuba.

WINB
P.O. Box 88, Red Lion, PA 17356
717 244 9316
Prime frequencies: 11790 kHz 1100–1400 UTC **NA SA**
11915 kHz 2100–0000 UTC **EU**
11950 kHz 0000–1100 UTC **NA SA**
12160 kHz 1900–2100 UTC **EU**
15715 kHz 1400–2400 UTC **AF EU**
Notes: After a political dispute between the owner and the general manager in the summer of 1995, WINB was pulled from the air and was scheduled to return a few months later.

WJCR
P.O. Box 91, Upton, KY 42784
Prime frequencies: 7490, 13595, 17595 kHz 24 hrs
Notes: Primarily broadcasts Christian country and gospel music.

WMLK
P.O. Box C, Bethel, PA 19507
Prime frequencies: 9465 kHz 0400–0700, 1700–2000 UTC **NA**
Notes: WMLK, operated by the Assemblies of Yahweh, is a conservative Christian broadcaster that is not frequently reported.

WRMI
8500 SW 8th St., Suite 252, Miami, FL 33144
305 267 1728
305 267 9253
71163, 11753 (CompuServe)
71163.1735@compuserve.com (Internet)
Frequency: 9955 kHz. Operates with various English- and Spanish-language programs on from 1100–0500 UTC.
Notes: WRMI broadcasts primarily Cuban clandestine and conservative Christian programs. The major programs on the station are La Voz de la Fundacion, Overcomer Ministries, Scream of the Chameleon, and Viva Miami.

WRNO
P.O. Box 100, New Orleans, LA 70181
504 889 0602
Prime frequencies: 7355 kHz 2300–0400 UTC **NA**
7395 kHz 0300–0600 UTC **NA**
15420 kHz 1400–2300 UTC **EU NA**
Notes: Originally a commercial broadcaster with some paid programming and plenty of relays of the WRNO-FM album

rock station, the station has become the relay outlet for many conservative and arguably facist political voices. The keystone of the programming has been Rush Limbaugh, but it also has featured the extremely right-wing politics of National Vanguard Radio, American Dissident Voices, and Ernst Zundel.

WSHB

P.O. Box 860, Boston, MA 02123

radio@csps.com (Internet)

Prime frequencies: 6095 kHz 1000–1300 UTC

7510 kHz 2000–0000 UTC

7535 kHz 0000–1000 UTC

21640 kHz 1600–2000 UTC

Also broadcasts in English on: 9370, 9430, 9455, 9840, 13770, 15665, 17510, and 17555 kHz.

Notes: WSHB is the North American branch of Monitor Radio International, the radio voice of the *Christian Science Monitor*.

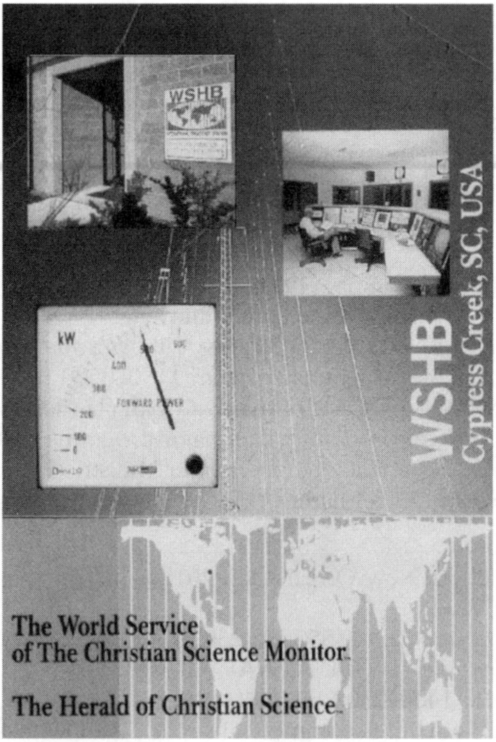

WSHB from the United States.

WVHA

P.O. Box 1844, Mount Dora, FL 32757

Prime frequencies: 7465 kHz 0200–0400 UTC **AF EU**

9852.5 kHz 2200–0200 UTC **AF EU**

9870 kHz 0700–1000 UTC **AS EU**

9930 kHz 1400–1600 UTC **AS**

11695 kHz 1200–1500 UTC **AS EU**

11930, 13770 kHz 1000–1100 UTC **AF EU**

13720 kHz 1800–2200 UTC **AS EU**

13770 kHz 1100–1200 UTC **AS EU**

15665 kHz 1500–1600 UTC **AS EU**

15745 kHz 1800–2000 UTC **AF**

17613 kHz 1700–1800 UTC **AF**

Notes: This station was formerly known as WCSN and was one of the Christian Science Monitor stations. The new owners, the Prophesy Countdown, air right-wing Christian end-times programming.

WWCR

1300 WWCR Ave., Nashville, TN 37218

Prime frequencies: 5065 kHz 2300–1400 UTC **AF AS EU NA**

5935 kHz 0100–1200 UTC **AF AS EU NA**

7435 kHz 0000–1100 UTC **AF AS EU NA**

9475 kHz 2100–2400 UTC **AF AS EU NA**

12160 kHz 1400–2300 UTC **AF AS EU NA**

13845 kHz 1200–0100 UTC **AF AS EU NA**

15685 kHz 1100–2100 UTC **AF AS EU NA**

Notes: Broadcasts a wide variety of programming, including a country music show, World of Radio, conservative politics (Radio Free America, Hour of the Time, For the People, etc.), Radio Newyork International, etc.

WYFR

290 Hegenberger Rd., Oakland, CA 94621

Prime frequencies: 5950 kHz 1000–1100 UTC **NA**

5950, 11830 kHz 1100–1400 **NA**

5985 kHz 0500–0700 UTC **NA**

6015, 11830, 17750 kHz 1400–1500 UTC **NA**

6085 kHz 0000–0100 UTC **NA**

6065, 9505 kHz 0100–0500 UTC **NA**

7355, 9985, 13695 kHz 0600–0700 UTC **AF EU**

9370 kHz 0400–0500 UTC **EU AF**

9985, 11580 kHz 0500–0600 UTC **AF EU**

11750, 11830, 17750 kHz 1500–1700 UTC **NA**
17845, 21525 kHz 2000–2100 UTC **AF EU**
21500, 21745 kHz 1600–1900 UTC **AF EU**

Notes: WYFR is the flagship of the large American Family Radio network, which broadcasts conservative Christian programming on AM, FM, and shortwave.

Uzbekistan

Radio Tashkent
72 Weedmore St., London W1H 9L, UK
Room 20, Numpoc Hotel, New Dehli 110001, INDIA

Prime frequencies: 7285, 9715, 15295, 17815 kHz 1200–1300, 1330–1400 UTC
5955, 5975, and 7285 kHz 0100–0130 UTC

Notes: Mail theft in Uzbekistan has been so common that several outside maildrops (including the two listed) have been announced for Radio Tashkent.

Vanuatu

Radio Vanuatu
P.O. Box 49, Port Vila

Prime frequencies: 3330, 3945, 4960, 6100, 7260 kHz 0900–1030 UTC

Notes: Radio Vanuatu is a small station from the South Pacific. Its broadcasts are only intended for a local audience. In addition to local programming, Radio Vanuatu also relays programming from some other shortwave stations, such as Radio Australia.

Vatican

Vatican Radio

Prime frequencies: 3945, 5882 kHz 2030–2050 UTC **EU**
3945, 6045 kHz 0600–0620 UTC **EU**
6095, 7305 kHz 0250–0315 UTC **NA SA**
6150, 7305, 9600, 11830 kHz 2245–2315 UTC **AS AUS**
7360, 9660 kHz 0320–0350 UTC **AF**
7360, 9660, 11625 kHz 0500–0530 UTC **AF**
9500, 11625, 15585 kHz 1345–1415 UTC **AS AUS**
9660, 11625, 13765 kHz 0630–0700 UTC **AF SA**
7355, 9645, 11625 kHz 2000–2030 **AF SA**

Notes: Vatican Radio is *the* source of information about the Catholic church. The broadcasts emanate directly from the Vatican and the Pope regularly addresses listeners.

Vatican Radio.

Venezuela

Radio National de Venezuela
Ap. 3979, Caracas
Prime frequency: 9540-kHz English program one hour per day.
Notes: The English program is rarely reported.

Vietnam

Voice of Vietnam
58 Quan Su St., Hanoi
84 4 255669
Prime frequencies: 5940 kHz 0400–0600 UTC
7285, 9732 kHz 1100–1130 UTC
9840, 12020, 15010 kHz 1000–1030 UTC
9840 kHz 1200–1230, 1330–1400, 1600–1630, 1800–1830, 1900–1930, 2330–2400 UTC

Alternate frequencies to 9840 kHz are 12020 and 15010 kHz.

Notes: The English Service of the Voice of Vietnam had not yielded strong signals to much of the world for years. But the Voice of Vietnam received a real boost in 1994 after Russia's old Radio Moscow transmitters became available for airtime.

Zambia

Radio Christian Voice
Private Bag E-606, Lusaka
260 1 274251
Prime frequencies: 4965 kHz 1500–2030 UTC
6065 kHz 1300–1500 UTC

Notes: Radio Christian Voice is a relatively new station on shortwave, having hit the airwaves at the end of 1994.

Zambia National Broadcasting Corporation
P.O. Box 50015, Lusaka
260 1 220864 74
Prime frequency: 6165, 7234 kHz 0300–2200 UTC

Notes: The ZNBC is well heard in Africa, but the signal pattern is much smaller than that of the new Radio Christian Voice.

Zimbabwe

Zimbabwe Broadcasting Corporation
P.O. Box HG 444, Highlands, Harare
263 4 795698
263 4 795698
Prime frequencies: 3306 kHz 2000–2200 UTC
4828 kHz 0300–0800 UTC

Notes: The ZBC broadcasts a steady schedule on various frequencies. To keep up with the English programs, check the

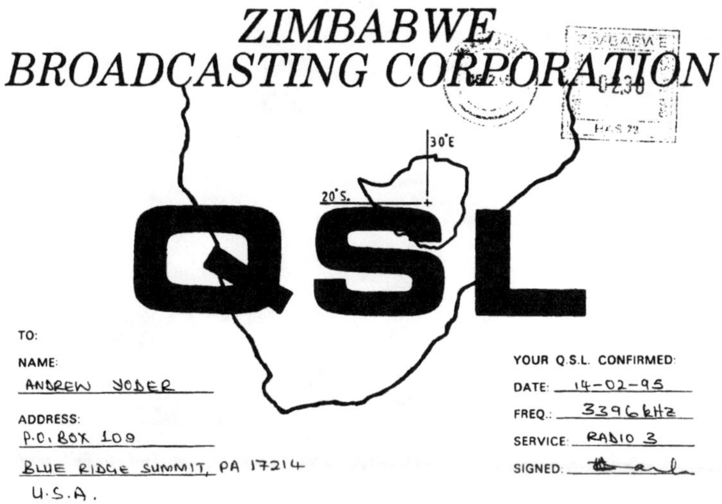

Zimbabwe Broadcasting Corporation.

other ZBC frequencies: 3396, 5012, 5975, 6045, and 7175 kHz. The ZBC broadcasts on shortwave are intended for local audiences and they sound much like a local AM station.

A

Making international calls and the UTC time chart

In the following list, the numbers beside the countries are the calling codes for that country and the numbers beside the city are codes for that city. For dialing instructions, check with the long-distance company in the region that you are calling from.

The two columns of numbers show you how to determine the local time from the universal time (UTC, formerly known as GMT). For example, see the listing for Albania. The left column (for standard time) lists +1 and the right column lists +2 (for summer time.) If you know what time it is UTC, then add one hour for the standard time in Albania or two hours for summer time. So, if it is 0000 UTC, it is 2 a.m. in Albania during the summer and 1 a.m. during the rest of the year.

Countries in the Western Hemisphere have negative numbers in the two columns. For example, Ecuador is listed as –5, –5. (The same number in both columns means that country doesn't use a daylight savings time plan.) If it is 0000 UTC, then it is 7 p.m. in Ecuador.

	Standard time	Summer time
Albania 355	+1	+2
Algeria* 213	+1	+1
Argentina 54	–3/–2	–4
Buenos Aires 1	–3	–4
Cordoba 51	–4	–4
Rosario 41	–3	–4
Ascension Islands 247	0	0
Australia 61	+8/9/9.5/10	+8/9.5/10/10.5/11
Melbourne 3	+10	+11
Sydney 2	+10	+11
Australian External Territory 672	+10	+11
Austria 43	+1	+2
Innsbruck 5222	+1	+2
Vienna 1	+1	+2
Azerbaijan 994	+4	+5
Bangladesh 880	+6	+6
Belgium 32	+1	+2
Antwerp 3	+1	+2
Brussels 2	+1	+2
Belize 501	–6	–6
Bhutan 975	+6	+6
Bosnia-Herzegovina 387	+1	+2
Botswana 267	+2	+2
Brazil 55	–2/3/4/5	–2/3/4
Brasilia 61	–2	–3
Rio de Janeiro 21	–2	–3
Sao Paulo 11	–2	–3
Bulgaria 359	+2	+3
Canada	–3.5/4/5/6/7/8	–2.5/3/4/5/6/7
China 86	+6/8	+7/9
Cook Islands 682	–10	–9.5
Costa Rica* 586	–6	–5
Croatia 385	+1	+2

	Standard time	Summer time
Czech Republic 42	+1	+2
Brno 5	+1	+2
Prague 2	+1	+2
Ecuador 593	–5	–5
Quito 2	–5	–5
Egypt 20	+2	+2
Alexandria 3	+2	+2
Cairo 2	+2	+2
Port Said 66	+2	+2
Estonia 372	+2	+3
Falkland Islands 500	–4	–4
Finland 358	+2	+3
Helsinki 0	+2	+3
France 33	+1	+2
Bordeaux 56	+1	+2
Marseille 91	+1	+2
Nice 93	+1	+2
Paris 1	+1	+2
Germany 49	+1	+2
Berlin 30	+1	+2
Bonn 228	+1	+2
Frankfurt 69	+1	+2
Munich 89	+1	+2
Dresden 51	+1	+2
Leipzig 41	+1	+2
Ghana 233	0	0
Greece 30	+2	+3
Athens 1	+2	+3
Rhodes 241	+2	+3
Guam* 671	+10	+10
Guatemala 502	–6	–5
Guatemala City 2	–6	–5
Honduras* 504	–6	–6
Hungary 36	+1	+2

	Standard time	**Summer time**
India 91	+5.5	+5.5
Iraq 964	+3	+4
Baghdad 121	+3	+4
Ireland 353	0	+1
Dublin 1	0	+1
Israel 972	+2	+3
Jerusalem 2	+2	+3
Tel Aviv 3	+2	+3
Italy 39	+1	+2
Florence 55	+1	+2
Naples 81	+1	+2
Rome 6	+1	+2
Venice 41	+1	+2
Japan 81	+9	+9
Hiroshima 82	+9	+9
Osaka 6	+9	+9
Tokyo 3	+9	+9
Yokohama 45	+9	+9
Kenya 254	+3	+3
Nairobi 2	+3	+3
Kuwait* 965	+3	+3
Latvia 371	+2	+3
Lebanon 961	+2	+3
Lesotho 266	+2	+2
Liberia* 231	0	0
Libya 218	+1	+2
Lithuania 370	+2	+3
Madagascar 261	+3	+3
Moldova 373	+2	+3
Monaco 33	+1	+2
all points 93	+1	+2
Mongolia 976	+8	+9
Myanmar 95	+6.5	+6.5
Namibia 264	+2	+2
Nepal 977	+5.75	+5.75

	Standard time	Summer time
Netherlands 31	+1	+2
Amsterdam 20	+1	+2
Rotterdam 10	+1	+2
The Hague 70	+1	+2
Netherlands Antilles 599	–4	–4
Bonaire 7	–4	–4
Curacao 9	–4	–4
New Zealand 64	+12	+13
Auckland 9	+12	+13
Christchurch 3	+12	+13
Wellington 4	+12	+13
Nigeria 234	+1	+1
Lagos 1	+1	+1
Norway 47	+1	+2
Bergen 5	+1	+2
Olso 2	+1	+2
Stavanger 4	+1	+2
Pakistan 92	+5	+5
Islambad 51	+5	+5
Palau 680	+9	+9
Papua New Guinea* 675	+10	+10
Phillipines 63	+8	+8
Manila 2	+8	+8
Poland 48	+1	+2
Portugal 351	0	+1
Lisbon 1	0	+1
Romania 40	+2	+3
Bucharest 0	+2	+3
Russia 7	+3–12	+4–13
Moscow 095	+3	+4
Rwanda 250	+2	+2
St. Helena 290	0	0
Saipan 670	+10	+10
Seychelles 248	+4	+4

	Standard time	Summer time
Sierra Leone 232	0	0
Singapore* 65	+8	+8
Slovakia 42	+1	+2
Bratislava 7	+1	+2
Kosice 95	+1	+2
Solomon Islands 677	+11	+11
South Africa 27	+2	+2
Capetown 21	+2	+2
Johannesburg 11	+2	+2
Pretoria 12	+2	+2
South Korea 82	+9	+10
Seoul 2	+9	+10
Spain 34	+1	+2
Barcelona 3	+1	+2
Madrid 1	+1	+2
Seville 54	+1	+2
Valencia 6	+1	+2
Sri Lanka 94	+5.5	+5.5
Colombo 1	+5.5	+5.5
Swaziland 268	+2	+2
Sweden 46	+1	+2
Goteborg 31	+1	+2
Stockholm 8	+1	+2
Swtizerland 41	+1	+2
Berne 31	+1	+2
Geneva 22	+1	+2
Lucerne 41	+1	+2
Zurich 1	+1	+2
Syria 963	+2	+3
Tanzania 255	+3	+3
Taiwan 886	+8	+8
Taipei 2	+8	+8
Thailand 66	+7	+7
Bankok 2	+7	+7

	Standard time	Summer time
Turkey 90	+2	+3
Ankara 4	+2	+3
Istanbul 1	+2	+3
Uganda 256	+3	+3
United Arab Emirates 971	+4	+4
Abu Dhabi 2	+4	+4
Dubai 4	+4	+4
United Kingdom 44	0	+1
Belfast 232	0	+1
Cardiff 222	0	+1
Edinburgh 31	0	+1
Glascow 41	0	+1
Liverpool 51	0	+1
London 71 and 81	0	+1
United States	–5/6/7/8	–4/5/6/7
Vatican City 39	+1	+2
all points 6	+1	+2
Venezuela 58	–4	–4
Caracas 2	–4	–4
Vietnam 84	+7	+7
Zambia 260	+2	+2
Zimbabwe 263	+2	+2

*No city code required

B

Guides to other shortwave services

Other than shortwave broadcast stations, which are covered within this book, many other books cover the other various services that are audible on shortwave. For further reference on these topics, especially for beginners, see:

American Radio Relay League (ARRL), *The ARRL Handbook for the Radio Amateur* (annual). Newington, Conn.: American Radio Relay League, Inc.

ARRL, *The ARRL Operating Manual.* Newington, Conn.: American Radio Relay League, Inc., 1991.

Bennett, Hank, David T. Hardy, and Andrew Yoder, *The Complete Shortwave Listener's Handbook, Fourth Edition.* Blue Ridge Summit, Pa.: TAB Books, 1994.

Dexter, Gerry L., *Shortwave Clandestine Confidential.* Reynoldsburg, Oh.: Universal Electronics, 1984.

Klingenfuss, J., *Guide to Radioteletype (RTTY) Stations.*

Kneitel, Tom, *National Directory of Survival Radio Frequencies.*

Kuperus, Bart, *WRTH Satellite Broadcasting Guide.* New York: Billboard, 1994.

Osterman, Fred, *The RTTY Listener.* Reynoldsburg, Oh.: Universal Electronics.

Passport to World Band Radio (annual). Penn's Park, Pa.: IBS.

Taggart, Ralph, *Weather Satellite Handbook.*

World Radio TV Handbook (annual). New York: Billboard.

Yoder, Andrew, *Pirate Radio: The Incredible Story of America's Underground, Illegal Broadcasters.* Solana Beach, Ca.: High-Text, 1996.

Zeller, George and Andrew Yoder, *The Pirate Radio Directory* (annual). Lake Geneva, Wisc.: Tiare Publications.

C

Power voltages and line frequencies (worldwide)

Afghanistan	AC 50/60, 220/380
Algeria	AC 50, 127/220, 220/380
Angola	AC 50, 220/380
Argentina	AC 50, 220/380
Australia	AC 50, 240/415
Austria	AC 50, 220/380
Bahrain	AC 50, 230/400
Bangladesh	AC 50, 230/400
Belgium	AC 50, 220/380
Bermuda	AC 60, 115/230
Bosnia	AC 50, 220/380
Bolivia	AC 50, 115/230
Brazil	AC 60, 127/220
Bulgaria	AC 50, 220/380
Canada	AC 60, 120/240
Chile	AC 50, 220/380
Colombia	AC 60, 120/240
Costa Rica	AC 60, 120/240
Croatia	AC 50, 220/380

Cuba	AC 60, 110/220
Cyprus	AC 50, 240/415
Czech Republic	AC 50, 220/380
Denmark	AC 50, 220/380
Dominican Republic	AC 60, 110/220
Ecuador	AC 60, 110/240
Egypt	AC 50, 220/380
El Salvador	AC 60, 115/230
Ethiopia	AC 50, 220/380
Fiji Islands	AC 50, 240/415
Finland	AC 50, 220/380
France	AC 50, 127/220, 220/380
Gabon	AC 50, 220/380
Germany	AC 50, 220/380
Ghana	AC 50, 220/400
Greece	AC 50, 220/380
Guadeloupe	AC 50, 220/380
Guam	AC 60, 110/220
Guatemala	AC 60, 120/220
Guyana	AC 50/60, 110/120
Holland	AC 50, 220/380
Hong Kong	AC 50, 200/346
Hungary	AC 50, 220/380
Iceland	AC 50, 220/380
India	AC 50, 230/400
Indonesia	AC 50, 127/220
Iran	AC 50, 220/380
Iraq	AC 50, 220/380
Ireland	AC 50, 220/380
Italy	AC 50, 220/380
Ivory Coast	AC 50, 220/380
Jamaica	AC 50, 110/220
Japan	AC 50/60, 100/200
Jordan	AC 50, 220/380
Kampuchea (Cambodia)	AC 50, 120/208
Kenya	AC 50, 240/415
Kuwait	AC 50, 240/415

Laos	AC 50, 220/380
Lebanon	AC 50, 110/190
Liberia	AC 60, 120/240
Libya	AC 50, 127/220
Luxembourg	AC 50, 120/208, 220/380
Malaysia (Sarawak)	AC 50, 240/415
Malta	AC 50, 240/415
Mauritius	AC 50, 230/400
Mexico	AC 60, 127/220
Monaco	AC 50, 127/220
Morocco	AC 50, 115/200
Mozambique	AC 50, 220/380
Myanmar	AC 50, 220/440
New Caledonia	AC 50, 220/380
New Guinea	AC 50, 127/220
New Zealand	AC 50, 230/400
Nigeria	AC 50, 230/415
Norway	AC 50, 230
Pakistan	AC 50, 230/400
Panama	AC 60, 110/220
Paraguay	AC 50, 220
People's Republic of China	AC 50, 220/380
Peru	AC 60, 220
Phillipines	AC 60, 110/220
Poland	AC 50, 220/380
Portugal	AC 50, 220/380
Puerto Rico	AC 60, 120/240
Romania	AC 50, 220/380
Russia	AC 50, 127/220
Saudi Arabia	AC 60, 110/220
Serbia	AC 50, 220/380
Singapore	AC 50, 230/400
Slovakia	AC 50, 220/380
South Africa	AC 50, 220/380
South Korea	AC 60, 100/200
Spain	AC 50, 127/220, 220/380
Sri Lanka	AC 50, 220/400

Sudan	AC 50, 240/415
Sweden	AC 50, 220/380
Switzerland	AC 50, 220/380
Syria	AC 50, 220/380
Tahiti	AC 50, 127/220
Taiwan	AC 60, 110/220
Tanzania	AC 50, 230/240
Thailand	AC 50, 220/380
Trinidad W. I.	AC 60, 115/220, 230/400
Turkey	AC 50, 220/380
Uganda	AC 50, 240/415
United Kingdom	AC 50, 240/415
United States	AC 60, 120/208, 120/240
Uruguay	AC 50, 220
Venezuela	AC 60, 120/240
Vietnam	AC 127/220, 220/380
Yemen	AC 50, 230/400
Zambia	AC 50, 220/380
Zimbabwe	AC 50, 220/380

D

Radio BBSs

While you are traveling, nothing (aside from shortwave) can help to keep you in touch with the world better than computer BBSs and networks. Some of the radio-related information that you can find on BBSs and computer networks includes schedules, loggings, background information and opinions about various programs, receiver reviews and opinions, and much more.

With a laptop computer and a modem, you can download plenty of current information. In fact, in many places, you can find a radio-related BBS within reach of a local telephone call. You can receive radio information via computer from three different sources: a radio network, a radio-related BBS, or a radio echo.

The *computer network* includes the Internet, America Online, CompuServe, Delphi, etc. These organizations have various radio forums, where listeners can post information, ask questions, and so on. The Internet radio forums are by far the most important of the networks, because nearly all of the major computer services (and many independent BBSs) include the Internet. Telephone numbers for the major services are not included here because you need to subscribe first. The various numbers can be supplied by the companies' information lines.

Radio echoes are much like the radio forums on the major computer networks—loads of posted information and questions. The difference is that these forums are carried over many independent BBSs. One of the largest of these in the world is the FidoNet. The FidoNet has hundreds of different forums, including several different radio forums. You can subscribe to these FidoNet forums—as a result, some BBSs carry all of the FidoNet forums and others only carry three or four. This chapter covers hundreds of different FidoNet BBSs.

Radio-related BBSs specialize in radio files and information. In addition to the same types of Q&A information that fills the radio echoes and radio forums on the major computer networks, radio-related BBSs also feature demo copies of different radio software, scanned photos of QSLs and radio artwork, radio shareware, etc. This appendix covers several dozen different radio-related BBSs. "Radio-related" in this case means that the BBSs feature either shortwave or amateur radio files.

Using BBSs

BBSs change regularly; different FidoNet forums are added and dropped, BBSs move and phone numbers change, and new BBSs open while others close. This chapter lists some of the many FidoNet and radio-related BBSs in existence around the world. Remember that some of this information will have surely changed by the time you read this.

If you are planning a trip in advance, be sure to check into a BBS to make sure that you can receive access. Some BBSs require extensive user information before they allow access; it's best to get this out of the way before making the trip. Also, some BBSs have subscription fees. If a particular BBS would be handy during a trip, but there is a fee, try contacting the BBS system operator and letting him or her know that you will be in town. The sysop might be willing to give you a temporary free or reduced-rate BBS membership.

If you plan to call BBSs for radio info while you're staying in a motel, be sure to check for fees on local calls. Most motels do charge for local calls, either per call or per minute, and the results can be quite expensive.

Radio-related Internet forums

alt.radio.digital
alt.radio.pirate
rec.radio.amateur.equipment
rec.radio.amateur.misc
rec.radio.broadcasting
rec.radio.info
rec.radio.noncom
rec.radio.shortwave

Radio-oriented BBSs

ANARC/ACE BBS	Kansas City, KS	913-345-1978
America Archive BBS	Cook County, IL	708-426-8903
American Silver Dollar	Alexandria, LA	318-443-0271
Bad to the Bone	Casco, MI	810-749-3581
Brian's Brainy BBS	Danville, IL	217-442-9818
Chi.-Area Comp. Center	Bolingbrook, IL	708-230-9068
FBN BBS	Champaign, IL	217-359-2874
The File Bank	Denver, CO	303-534-4646
The /Flux/ Line	Bloomfield Hills, MI	810-851-3509
Free Radio Network	Joplin, MO	417-624-1809
Grove BBS	Brasstown, NC	704-837-9200
HAM>link< RBBS	St. Paul, MN	612-HAM-0000
Ham It Up Data Systems	Moriches, NY	516-878-4906
KD3BJ Usenet BBS	Sellersville, PA	215-257-2744
Microcomputer Tech. BBS	Harrison, AR	501-365-7392
NAMU	Topeka, KS	913-273-1550
Radio Active	Atlanta, GA	404-437-1555
Radio Daze BBS	South Bend, IN	219-257-2138
The RF Deck	Baraboo, WI	608-356-4777
Saguaro Station	Phoenix, AZ	602-846-2318
Trance BBS	Torino, Italy	+39-11-482-751
WB3FFV Radio BBS	Baltimore, MD	410-661-2475

FidoNet BBSs

The following list contains some of the FidoNet-member BBSs from around the world. BBSs marked with an asterisk (*) are verified as a carrier for the FidoNet Shortwave Echo (SW Echo) forum. Those not marked with an asterisk might carry the SW Echo or some of the other radio-related FidoNet forums (Ham Forsale, Ham-Tech, etc.). If no radio-related files are carried by the FidoNet member BBS, it is possible that they will be picked up later on or that you might be able to convince the sysop to join.

Alabama

The Anchor Inn	Mobile	334-675-8406
The Computrion BBS	Birmingham	205-595-0183
*The Dateline BBS	Cullman	205-747-4194
The Edge of Cyberspace	Albertville	205-891-3403
*The Jack of All BBS	Hartselle	205-773-2859

Alabama *continued*

Shoal Creek eXpress	Florence	205-757-3355
*3 Nodes	Huntsville	205-551-9004

Alaska

TC's Byte Bank	North Pole	907-488-3751
Scriptorium	Anchorage	907-248-5985

Arizona

CNV	Phoenix	602-488-0656
The Desert Reef	Tucson	602-624-6386
Sunwise	Phoenix	602-584-7395
Union Jack	Phoenix	602-274-9921
The Wild Side BBS	Phoenix	602-258-8351

Arkansas

The Deserted Island	Little Rock	501-224-1605
Global Gateways	Conway	501-329-7508
Ozark Mt. BBS	Damascus	501-335-7871

California

*The Cat's Meow	Union City	510-471-0603
Crossroads	Irvine	714-724-1041
*Fairfield BBS	Fairfield	707-426-3119
*Gandalf's	San Diego	619-466-9505
*The Electronic Grapevine	Napa	707-257-2338
Happy Trails BBS	Orange	714-547-0719
Manetheran BBS	Irvine	714-509-9276
*9-1-1 BBS	San Diego	619-669-0385
Online Resource	San Diego	619-793-8360
Silver Cactus BBS	Lancaster	805-949-7703
Sleuth BBS	Chatsworth	818-727-7639
The Solar System BBS	Mission Viejo	714-837-9677
*Ursa Major	Manhattan Beach	310-545-5611
*Zooming	San Diego	619-277-4140

Colorado

Access Denied	Arvada	303-431-6796
Colorado Connection	Arvada	303-423-9775

The Dart Board	Dolores	303-882-2360
Dimensions Online	Aurora	303-363-0910
*The File Bank, Inc.	Denver	303-534-4646
The Forum	Ft. Collins	303-226-4218
Grotto Lounge	Arvada	303-421-6965

Connecticut

Applause BBS	Waterbury	203-754-9598
Erasmus BBS	Newington	203-666-5113
*Hatters Park	Danbury	203-744-0179
*The Oracle	Bethel	203-794-8675
Shotgun BBS	Stamford	203-969-0825
USS Enterprise	Monroe	203-261-6863

Delaware

Baseline BBS	Newark	302-834-1089
The DVUG BBS	New Castle	302-324-8091
Imperialink BBS	Wilmington	302-892-9953

Florida

*Bullwinkle's Corner	Orlando	407-896-5772
Civil Air Patrol	Bradenton	813-750-9051
*Cornucopia TBBS	Winter Park	407-645-4929
CyberNexus	Merritt Island	407-459-9100
Esoteric Oracle	Gainesville	904-332-9547
Field of Dreams	Jacksonville	904-241-0735
*The Firehouse BBS	Lake Buena Vista	904-934-8678
Genesis BBS	Margate	305-753-5033
*Ground Zero	St. Petersburg	813-849-4034
*LASER!	Orlando	407-647-0031
Last Days	Mt. Dora	904-735-0531
Longbow BBS	Tampa	813-961-3653
*Mercury Opus	St. Petersburg	813-321-0734
Neon Knight's BBS	Crestview	904-689-3692
Nightmare Cafe	Panama City	904-874-2296
Parados	Jacksonville	904-260-3172
Prime Time BBS	Inverness	904-637-3713
Ups & Downs	Cooper City	305-434-8403
USS Enterprise EBBS	Englewood	813-473-1246
Whistler's Hollow	Jacksonville	904-727-9289

Georgia

American Connection	Savannah	912-927-7323
Crack of Dawn BBS	Ft. Stewart	912-369-7023
Deep-Sky BBS	Atlanta	404-321-5904
The Hotel California	LaGrange	706-845-7102
Planet X BBS	Gainesville	404-536-1652
*Total Recall	Douglasville	770-920-0621

Hawaii

Color Computer Library	Honolulu	808-545-8368

Idaho

*The Bodhi Tree	Boise	208-327-9916

Illinois

Boomtown BBS	McConnell	815-868-2422
*CfC	Chicago	708-362-7875
Creative Thoughts BBS	Barrington	708-382-3904
Dark Elf BBS	Cary	708-516-8953
East Village BBS	Chicago	312-777-2574
Gateway Elite BBS	Rockford	815-398-4678
Hook & Slice Club BBS	Danville	217-446-0105
*JPUSA BBS	Chicago	312-878-6030
National Islamic	Chicago	312-274-8136
RuneQuest	Chicago	312-728-7784
Shadowgate BBS	Sterling	815-622-9639
Speedy's Dilemma	Murphysboro	618-687-4806
Squirrel's Nest	Marseilles	815-795-6371
Steve's Maildrop & BBS	Columbia	618-281-8702
Uncle Bob's BBS	Lake Villa	708-265-0695
Under the Influence	Morris	815-942-2930
Village Oak BBS	Oak Lawn	708-430-7732

Indiana

ArcadiaVision BBS	Kouts	219-766-2378
Computer Support BBS	South Bend	219-272-8129
Electronic Warfare BBS	Vincennes	812-882-0644
4th Dimension BBS	Terre Haute	812-299-2983
The Flaming Star BBS	Terre Haute	812-898-2561
Graffiti o/t BBS Wall	Lafayette	219-448-2842
I.O. Board	Anderson	317-644-3039

Ivy Tech St. College	Terre Haute	812-299-9306
*Lakeland BBS	LaGrange	219-463-2484
The Software Zone	Martinsville	219-342-2094
The Sometimes BBS	Portage	219-763-2031

Iowa

Helium High-Grounds	Ames	515-232-0969

Kansas

*ANARC/ACE BBS	Kansas City	913-345-1978
The Quelle	Kansas City	913-287-2600

Kentucky

The Barbarian's Hut	Ft. Knox	502-942-8046
The ScrapYard	Ft. Knox	502-942-0864
U.S. Aviators Domain	Erlanger	502-342-6554
VACIS	Worthington	606-836-1267

Louisiana

Black Boar Inn	Metairie	504-885-3960
*The Digital Cottage	New Orleans	504-897-6614
Heavy Metal BBS	Houma	504-872-3043
N.Louisiana Connect	Ruston	318-255-4497
Southern Star BBS	New Orleans	504-885-5928

Maine

The Wizards Guild	Lewiston	207-783-0874

Maryland

*≡ACE≡ ONLINE	Silver Spring	301-942-2218
Bifrost	Mt. Ranier	301-779-9381
Econo-RBBS	Catonsville	410-747-3619
The Emerald Mines BBS	Largo	301-499-0326
G-Comm	Warren	410-436-7638
*Hafa Adai Exchange	Great Mills	301-994-9460
*The Last DX Connection	Bowie	301-805-8921
*MetroNet	Columbia	410-720-5506
MirageMecca RBBS	Baltimore	410-426-5097
*NFB Net	Baltimore	410-752-5011
Ron's Room	Severn	410-969-2241

Massachussetts

Computer Castle Link	Haverhill	508-521-6941
*Tom's BBS	Braintree	617-356-3538
*World of Burgers	Worcester	508-753-6969

Michigan

*The Air Studio BBS	Livonia	313-522-5349
Chess Players Forum	Lincoln Park	313-386-7054
Eau Claire Connection	Eau Claire	616-461-6801
The Programmer's Edge	Livonia	810-477-6695
The Rainbow Bridge BBS	Coldwater	517-278-7029
The Raven!	Redford	313-937-2016

Minnesota

The Black Hole BBS	Chaska	612-442-5682
*HAM>link< RBBS	Saint Paul	612-HAM-0000
Infiniti	Richfield	612-861-7460
SpareCom	Shakopee	612-445-5755

Mississippi

After Hours BBS	Jackson	601-371-0423
The Eagle's Eye	Vicksburg	601-636-3212
The Gateway	Biloxi	601-374-2697
Mid-South BBS	Amory	601-256-1494
StarNet BBS	Jackson	601-981-9228

Missouri

The Batboard TBBS	Columbia	314-446-0475
B.S. Box	St. Louis	314-861-3902
*Cork's Place BBS	Joplin	417-659-8666
*Hard Rock BBS	Nixa	417-725-8003
The Lake BBS	Osage Beach	314-348-9050
Mega-Link BBS	St. Joseph	816-364-1035
*Sound Advice	Gladstone	816-436-4516
*St. Louis Online	St. Louis	314-561-4956
*Too Tall'S	Ladue	314-997-7060
*The Village Crier	St. Joseph	816-387-8881

Nebraska

Castle Keep BBS	Bellevue	402-292-0789
MacNet Omaha	Elkhorn	402-289-2899

Nevada

ICON BBS	Las Vegas	702-256-4107

New Hampshire

Botany Bay	Portsmouth	603-431-7090
The Cereal Port BBS	Rindge	603-899-3335
The Computer Castle	Newton	603-642-5949
EMail Connection	Rochester	603-335-0213
The Outpost	Nashua	603-888-3840

New Jersey

Alchemiga	Sparta	201-398-6360
Altered Illusions	Dayton	908-329-3216
The Armory BBS	Phillipsburg	908-859-0162
Beacon Studios BBS	Union City	201-863-5253
Beyond Reality	Fords	908-417-2175
The Bitter End BBS	Browns Mills	609-893-2874
Blackstar	Boonton	201-335-6132
The Black Tower	Rockaway	201-361-6651
The Boss BBS	Tenafly	201-568-7293
Dmaster's Den	Madison	201-301-9583
The Flight Deck	Ft. Monmouth	908-389-3202
The GearBox	New Milford	201-692-1110
Hard Drive Cafe BBS	Wayne	201-790-6300
Harry's Place	Mahwah	201-934-0861
HQ Informational Service	Orange	201-672-8969
Instant Access	Newark	201-817-9111
In the Wind BBS	Hackensack	201-646-0227
It's All Rock'n'Roll	Trenton	609-695-9319
Just Programs BBS	Roselle Park	908-298-9098
The Licking Factory	Colonia	908-815-3146
MicroFone	Metuchen	908-494-8666
Mt. HED BBS	Parsippany	201-625-1519
Nightlink BBS	Manahawkin	609-597-4290
The Null Pointer BBS	Madison	201-301-2182
The Perverts BBS	Union Beach	908-888-0176
*Pics Online Multiuser System		609-753-2540
*Plain Brown Wrapper BBS	Cresskill	201-569-6685
Planet Shadowstar TBBS	Edison	908-494-3417
The Roach Motel	Nutley	201-667-3326
Shockwave Rider	Freehold	908-294-0659

New Jersey *continued*

The Sound Connection	Scotch Plains	908-322-0131
S.Coast Online Services	Ocean City	609-399-5708
The Tammy Board BBS	Mount Hope	201-361-5954
Union Lake BBS	Millville	609-327-5553
The Vortex BBS	Belleville	201-751-5608
The Waterside BBS	Little Ferry	201-641-5375

New Mexico

Construction Net #6	Los Alamos	505-662-0659
The Dragon's Lair	Portales	505-359-1770
Selective Online	Sante Fe	505-473-9765
Tel-Us Computer BBS	Las Vegas	505-425-6995

New York

ABC Online	Queens	718-446-2157
The Aquarium/AWE BBS	Buffalo	716-885-8363
The Belfry	New York	718-793-4796
The Boiler Room	Brooklyn	718-265-2532
The Brewster BBS	Brewster	914-279-2514
Coffee Clutch BBS	Clifton Park	518-383-3156
Control Central	Islandia	516-342-0382
The DarkLands BBS	Herkimer	315-866-8187
Dark Side of the Moon	Crompond	914-736-3186
Dragon's Cove	Rome	315-339-0208
East End Computer	Manorville	516-395-1033
The Enigma	Oswego	315-343-6706
*Ham-Net BBS	Manlius	315-682-1824
Harbringer BBS	Holbrook	516-472-1036
The Hidden Pyramid	East Syracuse	315-433-5398
Holman's World	New York	718-529-8890
HTBBS Network Services	Kew Gardens	718-380-8003
Kraut Haus	Rochester	716-359-0871
The Link BBS	Staatsburg	914-889-8379
The Machine BBS	Modena	914-883-6612
Medina Online	Medina	716-798-5549
Meng's Madhouse	Hilton	716-964-8323
The Messed BBS	Freeville	607-844-9216
MicroQuick	Plainview	516-935-5704
*The Night Owl BBS	Saratoga Springs	518-581-1797
The Pier BBS	Brooklyn	718-253-3583

The Promised Land	Bronx	718-295-3266
Quantum 2000 BBS	Queens	718-740-8411
Real Exposure	New York	212-691-2679
*Red Onion, ExPress	Wawayanda	914-342-4585
SMBBS Network	Wappingers Falls	914-297-2915
Tree Branch	Briarwood	718-739-5845
The United Front BBS	Flushing	718-886-6797
*Unity BBS!	Syracuse	315-488-0679

North Carolina

ANSI-Mation Alley	Jacksonville	910-346-6543
*Borderline! BBS	Concord	704-792-9241
Free Advice	Smithfield	919-934-1002
Hawaiian Hang Time	Hubert	910-326-5098
Hobbies Unlimited	Raleigh	919-821-4354
The Leap BBS	Cerro Gordo	910-654-5593
*The Mach] [BBs	Thomasville	910-475-5662
The Maniac's Den	Carthage	910-947-3077
*The North State BBS	Durham	910-220-0291
Patriot BBS	Graham	910-222-8524
The Pig Pen BBS	Fayetteville	910-487-0363
Terminal Entry BBS	Rockingham	910-895-0368
The Vault BBS	Raleigh	919-821-4049
Wilmington Online	Wilmington	910-763-3849

North Dakota

The D-Generation	Bismarck	701-258-0872

Ohio

Bordertown/TexMex	Woodville	216-862-3423
Bulldog's Lounge	Wheelersburg	614-574-6590
The Consciousness BBS	Sciotoville	614-776-2618
The Crystal Palace	Port Clinton	216-797-4719
Freedom BBS	Streetsboro	216-655-9626
The Mother Board Club	Buckeye Lake	614-928-2259
The Mountain Lair	Belpre	614-423-0567
The New Frontier	Akron	216-733-0979
*PC-OHIO	Cleveland	216-381-3320
Spock's Adventure BBS	Cincinnati	513-779-9717
Twisted Reality	North Olmsted	216-779-4113
*WLIO TBBS	Toledo	419-228-8227

Oklahoma

*The Ham Radio Emporium	Owasso	918-272-4327
Light Speed] [	Tulsa	918-299-4811
The Northern Exchange	Tulsa	918-747-2273
*Retriever's Retreat	Edmond	405-943-8638

Oregon

A bit of Aloha	Aloha	503-591-7542
*Chemeketa Online	Chemeketa	503-393-5580
Hell Above Ground	Portland	503-284-3569
The Inferno	Eagle Creek	503-637-3178
*NWCS Online	Oregon City	503-655-3927
Ultimate BBS	Eugene	503-461-1148

Pennsylvania

Alpha Beta	Pittsburgh	412-683-4327
The Annex BBS	Pittsburgh	412-635-9165
Antarctica BBS	York	717-755-2440
Bart & Homer's BBS	Wernersville	610-678-2431
*BlinkLink	Pittsburgh	412-766-0732
Cheers BBS	Johnstown	814-539-6648
The Circuit Board	Greentown	717-676-9339
Cybernetics BBS	Ephrata	717-738-1976
The DataWorks	Dickson City	717-489-0862
The Eagle's Lair	Bellwood	814-742-9364
Edge of Chaos	Philadelphia	215-564-4208
Freddies' Playhouse BBS	Tarentum	412-226-3129
Freeland BBS	Freeland	717-636-0131
Gateway to End of Time	Collegeville	610-454-9862
The Hotline BBS	Lansdale	215-393-8594
The Keep BBS	Lansdale	215-855-0401
Keystone Amiga BBS	Allentown	610-770-0774
Memory Prime BBS	Waynesburg	412-627-3227
Milliways	Pittsburgh	412-766-1086
The Northeast Filebank	Jermyn	717-876-0152
Operation Mindcrime	Altoona	814-742-9885
Pennsylvania Online	Harrisburg	717-657-8699
Pirate's Den BBS	New Castle	412-652-1694
RBAS Unlimited	Trevose	215-357-8177
The Solution II	Richlandtown	215-529-9501
Tim's BBS	Erie	814-825-8660

Rhode Island

*Dimensions	Westerly	401-596-3502

South Carolina

Chaos	Sumpter	803-469-9267
Charleston Police Dpt.	Charleston	803-763-0846
Chris & Nancy's Place	Sumpter	803-494-2082
Custom Computers BBS	Anderson	803-375-0074

South Dakota

*Ducky's	Rapid City	605-393-9626

Tennessee

*The GOOD News BBS	Chattanooga	615-698-0407
Joe's Garage BBS	Bartlett	901-382-0268
The Hotel California	Johnson City	615-928-5704
*Music City Archives	Nashville	615-952-2254
*The Nashville Exchange BBS	Nashville	615-383-0727
The POST BBS	Knoxville	615-691-1887
The Shoreline BBS	Memphis	901-775-3190
Sam's Smalltown Diner	Greenbrier	615-643-7562
The Tea Room BBS	Maryville	615-681-2387
The Vision BBS	Murfreesboro	615-896-7949

Texas

The Asylum BBS	San Antonio	210-637-5670
The Black Gate	El Paso	915-585-3701
Botany Bay	Tyler	903-509-8518
C-Link Online Services	Desoto	214-223-8363
*Dallas Remote Imaging Group BBS	Carrollton	214-394-7438
The Digital Outlet	Borger	806-273-2834
Ft. Worth Online	Ft. Worth	817-735-8166
JAXBBS	Dallas	214-823-1579
Loengrube Mailbox	Odessa	915-550-5122
MacEndeavor	Houston	713-640-1298
Neon Jungle	San Antonio	210-656-0109
Pawn to King's Four	Huntsville	409-291-3322
Poseidon	El Paso	915-593-0639
*Programmer's Connection	Longview	903-643-7607

Texas *continued*

The Raster Line	Houston	713-568-4128
Recompression Chamber	Brownfield	806-637-8113
*Resonant Frequency	Mesquite	214-686-0427
*The RF-BaKer	Borger	806-273-2407
Savage Jungle	Hurst	817-268-1914
*User-To-User	Dallas	214-393-9317
*WBBS	Conroe/Houston	409-447-4267
Ziggyuken	Cleburne	817-558-4600

Utah

The Cedar Chest BBS	Cedar City	801-586-8751
The Iron Grid	Salt Lake City	801-486-0929
Love Data	Syracuse	801-776-3459
South Valley BBS	West Jordan	801-567-0036

Virginia

The COMA BBS	Charlottesville	804-293-2400
The Elusive Diamond	Burke	703-323-6423
The Genesys BBS	Virginia Beach	840-499-9101
Kerry's Place	Moneta	540-297-8458
Mid-Atlantic OS/2 Group	Virginia Beach	804-422-8462
*Northern VA Astronomy Club	Falls Church	703-256-4777
Shareware Solutions BBS	Virginia Beach	804-490-9630
Sparkies Machine BBS	Roanoke	540-366-4299
Viginia Data Exchange	Newport News	804-877-3539
Virginia Shareware Net	Dale City	703-730-8731
Woody's Warehouse	Woodbridge	703-878-3664

Washington

Cameron's Railroad	Arlington	206-659-2132
Columbia Basin BBS	Moses Lake	509-766-2867
The Dark Masters' BBS	Spanaway	206-846-8312
Dino's Doghouse	Oak Harbor	206-679-6971
Extensions BBS	E. Wenatchee	509-886-0306
The Final Frontier	Everett	206-303-9153
Four Aces	LaConner	206-466-ACES
More than Meets the Eye	Everett	206-787-5339
Night Voyager BBS	Veradale	509-926-1686
NW Disability Access	Tacoma	206-539-0704

*NorthWest Online	Spokane	509-244-DOOM
OS/2 Northwest Group	Bellevue	206-957-4513
Search BBS	Vancouver	206-253-5213
Starwest BBS	Clarkston	509-758-6248
Wally World BBS	Walla Walla	509-529-3726

West Virginia

The Bit Bank	Charlestown	304-728-0884
The Black Hole	St. Albans	304-727-5711
Coal County BBS	Hurricane	304-562-2263
The Empire BBS	Kincaid	304-465-5223
Killarney Narrows	Huntington	304-523-8643
The Mindless Ones	Weirton	304-723-2133
Seneca Station	Elkins	304-636-9592
Starbase 90 BBS	Crab Orchard	304-252-6390

Wisconsin

*Aarrgghh! BBSq	West Bend	414-334-7041
The Decker's BBS	Eau Claire	715-839-0942
*Exec-PC BBS	Milwaukee	414-789-4360
*Radio Free Milwaukee	Milwaukee	414-351-1823
The Shire	Sparta	608-269-7019

Australia

*Ancient Evenings	Perth	+61-9-250-7533
*The Christian BBS	Sunshine Coast	+61-074-466-388
*Computer Connection	Adelaide	+61-08-326-2388
*Cross Facts BBS	Russell Lea	+61-2-712-3910
*First Release Shareware		+61-03-879-9160
*Fox's Lair BBS	Brisbane	+61-7-380-36821
*Heaven's Door	Sydney	+61-2-415-6079
*Hot-Line BBS	Adelaide	+61-8-373-5136
*The InterACTive BBS	Canberra	+61-06-239-7311
*Melbourne PC User Group	Melbourne	+61-3-699-6788
*Oracle PC-Network	Adelaide	+61-08-234-0791
*The OS/2 Cellar BBS		+61-7-216-2539
*Phoenix BBS	Sydney	+61-02-585-1925
*Power Up Oz		+61-7-399-1322
*The Radio Shack BBS	Melbourne	+61-3-9532-5737
*SATCOM_Australia	Sydney	+61-2-905-0849

Australia *continued*

*Short Circuit BBS	Sydney	+61-02-569-6259
*~Shortwave Possums BBS~	Sydney	+61-2-651-3055
*Soft-Tech	Queensland	+61-7-869-1311
*The Twilight Zone	Sydney	+61-2-750-6117
*Vulcan's World		+61-02-635-1204
*500ccFI TBBS		+61-2-565-1630

Belgium

*DXA-<Shortwave>-BBS	Antwerp	+32-3-8253613

Brazil

Sherwood BBS	Sao Paulo	+55-011-889-9677

Canada

*ABS International	Montreal	514-937-7451
*Astron BBS	St. John	506-652-8999
Calanost BBS/2	Edmonton	403-468-1741
*Channel-23	Orleans	613-830-8147
CompuBBS	Edmundston	506-735-3831
*Coven's Den	Vanier	613-746-3584
*CRS Online	Toronto	416-467-4975
Eco Communications BBS	Winnipeg	204-253-6711
*Freddy's Place	Edmonton	403-456-4241
*.\\iramax BBS	Vancouver	604-323-9698
*One Stop BBS	Vanier	613-746-3584
*The ODXA Listening Post	Aurora	905-841-6490
Pro Link BBS	Amos	819-732-7710
Quebec Online	Montreal	514-486-8959
SAUG	Saskatoon	306-242-6936
*SLIP	Alberta	403-299-9900
*Space Lodge BBS	Bowmanville	905-697-2935
The Staff Room BBS	Tecumseh	519-979-4208
TOTSE/2	Spencerville	613-658-5331
*Voice of Windsor BBS	Windsor	519-969-4292

France

*Minidou BBS	Ra Conte	+33-47-679189

Mexico

Pasito Tun Tun	Monterrey	+52-8-356-8446

The Netherlands

*Alexander BBS	Rotterdam	+31-10-4565600/2201454
*AINEX-BBS		+31-33-633916/653078
McBaud! BBS	Vlaardingen	+31-10-475-2961
*Omnix BBS		+31-0-1680-26819
*Scoop BBS		+31-3499-96366

New Zealand

*SunCity BBS	Nelson	+64-03-548-9171
*TV Mania	Christchurch	+64-3-352-8586
*WAG BBS	Foxton Beach	+64-6-363-6015

Puerto Rico

The Glass BBS	Mayaguez	809-832-4984
The Tropical Island BBS	Hormigueros	809-849-5921

Sweden

*GET	Lidingo	+46-8-7655670
*SweDX (Swedish DX Federation)	Malmoe	+46-853034727

United Kingdom

*The Control Center		+44-161-7079221
*The Crooked Spire Mult-BBS		+44-1246-551626/554626
*Pauls Point		+44-0191-4569135
*The Radio Shack		+44-1947-897551
*Golly! on Shortwave	Twyford	+44-1734-320812
The Rock of Gibraltar		+44-0181-6786087

Index

Illustration page numbers are in **boldface**.

F

G

H

I

J

K

L

R

Y

Z

About the author

At the tender age of 13, Andrew Yoder became addicted to shortwave listening on a World War II era Hallicrafters SX-28A in the basement of one of his best friends. Since that time, he has spent numerous hours glued to radios, in search of shortwave and pirate broadcast stations. Andrew has been bitten by the nostalgia bug and currently owns a small lineup of tube radios from the 1940s through the 1960s. His articles have appeared in *Popular Communications, Popular Electronics, Radio!*, and *New Jersey Monthly*, and he writes regular columns for several newsletters in North America and Europe. His recent books include *Pirate Radio: The Incredible Saga of America's Underground Illegal Broadcasters, Auto Audio: Choosing, Installing, & Maintaining Car Stereo Systems,* and *Build Your Own Shortwave Antennas*.